Microarrays & Microplates

Applications in Biomedical Sciences

Microarrays & Microplates
Applications in Biomedical Sciences

S. Ye & I.N.M. Day (Eds)
School of Medicine, University of Southampton, Southampton, UK

First published 2003

A CIP catalogue record for this book is available from the British Library.

ISBN 1 85996 074 X

BIOS Scientific Publishers Ltd
9 Newtec Place, Magdalen Road, Oxford OX4 1RE, UK
Tel. +44 (0)1865 726286. Fax +44 (0)1865 246823
World Wide Web home page: http://www.bios.co.uk/

Distributed exclusively in the United States of America, its dependent territories and Canada, Mexico, Central and South America, and the Caribbean by Springer-Verlag New York Inc., 175 Fifth Avenue, New York, NY 10010-7858, by arrangement with BIOS Scientific Publishers Ltd, 9 Newtec Place, Magdalen Road, Oxford OX4 1RE, UK.

Production Editor: Phil Dines
Typeset by Charon Tec Pvt. Ltd, Chennai, India
Printed by TJ International Ltd., Padstow, UK

Contents

Colour plates can be found between pages 22 and 23.

Contributors

M. A. Al-Dahmesh, Human Genetics Division, School of Medicine, Southampton University Hospitals NHS Trust, Southampton, UK

K. K. Alharbi, Human Genetics Division, School of Medicine, Southampton University Hospitals NHS Trust, Southampton, UK

Julian F. Burke, Genetix Limited, New Milton, UK

X. Chen, Human Genetics Division, School of Medicine, Southampton University Hospitals NHS Trust, Southampton, UK

Ian N. M. Day, Human Genetics Division, School of Medicine, Southampton University Hospitals NHS Trust, Southampton, UK

P. J. R. Day, Wellcome Trust Centre for Human Genetics, University of Oxford, Oxford, UK

Per Eriksson, Atherosclerosis Research Unit, King Gustav V Research Institute, Karolinska Institute, Stockholm, Sweden

R. H. Ganderton, Human Genetics Division, School of Medicine, Southampton University Hospitals NHS Trust, Southampton, UK

T. R. Gaunt, Human Genetics Division, School of Medicine, Southampton University Hospitals NHS Trust, Southampton, UK

Susan Henshall, Cancer Research Programme, Garvan Institute of Medical Research, Sydney, Australia

L. J. Hinks, Human Genetics Division, School of Medicine, Southampton University Hospitals NHS Trust, Southampton, UK

M. R. James, Wellcome Trust Centre for Human Genetics, University of Oxford, Oxford, UK

Masato Mitsuhashi, Department of Pathology, University of California, Irvine, USA

Taku Murakami, Hitachi Chemical Research Center, University of California, Irvine, USA

C. David O'Connor, Division of Biochemistry and Molecular Biology, University of Southampton, Southampton, UK

S. D. O'Dell, Human Genetics Division, School of Medicine, Southampton University Hospitals NHS Trust, Southampton, UK

Tomi Pastinen, Montreal Genome Centre, Montreal, Quebec, Canada

Karen Pickard, Division of Biochemistry and Molecular Biology, University of Southampton, Southampton, UK

Lawrence M. Schwartz, Biophysics Program, Harvard University, Cambridge, USA

Laurence Shumway, Department of Biology, University of Massachusetts, Amherst, USA

E. Spanakis, Human Genetics Division, School of Medicine, Southampton University Hospitals NHS Trust, Southampton, UK

Sarah K. Stephens, Genetix Limited, New Milton, UK

M. A. Suchard, Wellcome Trust Centre for Human Genetics, University of Oxford, Oxford, UK

Mark R. Truesdale, Genetix Limited, New Milton, UK

Carl Whatling, Atherosclerosis Research Unit, King Gustav V Research Institute, Karolinska Institute, Stockholm, Sweden

Shu Ye, Human Genetics Research Division, School of Medicine, University of Southampton, Southampton, UK

B. B. Zhang, Wellcome Trust Centre for Human Genetics, University of Oxford, Oxford, UK

Abbreviations

ABTS	2,2-azino-di(3-ethylbenzene-thiazoline-6-sulfonic acid)
AGPC	acid-guanidine/phenol/chloroform
APTS	3-aminopropyl-triethoxysilane
ARMS	amplification refractory mutation system
Aβ	amyloid beta
BCIP/NBT	5-bromo-4-chloro-3-indolyl phosphate/nitoblue tetrazolium
bDNA	branched DNA
BrdU	bromo-deoxy-uracil
BSA-NHS	bovine serum albumin-*N*-hydroxysuccinimide
CCD	charge coupled device
CEA	carcinoembryonic antigen
CEPH	Centre d'Etude du Polymorphisme Humain
CFTR	cystic fibrosis transmembrane conductance regulator gene
CPT	cycling probe technology
DAB	3,3′-diaminobenzidene
DGGE	denaturing gradient gel electrophoresis
DHPLC	denaturing high-performance liquid chromatography
DIG	digoxigenin
DMF	dimethyl formamide
DMSO	dimethylsulfoxide
ECL	enhanced chemiluminescence
ELISA	enzyme-linked immunosorbent assay
ER	oestrogen receptor
EST	expressed sequence tag
EtOH	absolute alcohol
FISH	fluorescence *in situ* hybridization
FKBP	FK506 binding protein
FMOC	9-fluorenylmethyloxycarbonyl
FRAP	FKBP-rapamycin associated protein
FRP	FKBP-12-rapamycin binding domain
GAPDH	glyceraldehydes-3-phosphate dehydrogenase
GFP	green fluorescent protein
GQA	general purpose query application
GST	glutathione-*S*-transferase
GTPS	3-glycidoxypropyl-trimethoxysilane
H&E	haematoxylin and eosin
HEPA	high efficiency particulate air
His_6	hexa-histidine
IC	immunochemistry
ICC	immunocytological chemistry
IHC	immunohistochemistry
IL	interleukin
ISH	*in situ* hybridization
LIMS	laboratory information management systems
MADGE	microplate array diagonal gel electrophoresis
MAGIChip	microarray of gel-immobilized compounds on a chip
MALDI	matrix-assisted laser desorption ionization
MAP	mitogen-activated kinase
MBP	mannose-binding proteins
MeOH	methanol
MS	mass spectrometry
NASBA	nucleic acid sequence-based amplification
NBA	normal body atlas

NVOC nitroveratryloxycarbonyl
ORF open reading frame
OTMS *n*-octadecylmethoxysilane
PAGE polyacrylamide gel electrophoresis
PCR polymerase chain reaction
PDITC 1,4-phenylene diisothiocyanate
PDMS poly(dimethylsiloxane)
PI phosphoinositides
PIN prostatic intraepithelial neoplasia
PMP paramagnetic particle
PNA peptide nucleic acid
PR progesterone receptor
PSA prostate-specific antigen
PTM post-translational modification
PVDF poly(vinylidene difluoride)
RDA representational differences analysis
RPA ribonuclease assay
RT-PCR reverse transcription–polymerase chain reaction
SAGE serial analysis of gene expression
SAM self-assembled monolayer
SBH sequencing by hybridization
scFv single chain variable fragment
SDA strand displacement amplification
SELDI-MS surface-enhanced laser desorption/ionization mass spectrometry
SNP single nucleotide polymorphisms
SPR surface plasmon resonance
SQL Structured Query Language
TAG-SBE tagged single base pair extension
TGGE temperature gradient gel electrophoresis
TMA tissue microarray
TTGE temporal thermal gradient electrophoresis
UPA universal protein array

Preface

Microarray analysis is one of the most significant biotechnological developments in recent years. The advent of this technology is synchronous with the rapid progress of the Human Genome Project. This is clearly not a coincidence; instead, there is a logical connection. The availability of the large amount of genomic sequence data generated by the Human Genome Project empowers microarray analysis, while the utility of this technique for systematic analysis of gene expression/function and for high throughput analysis of genetic variants makes it a very important tool in the 'post-genome' era.

Most microarray-based methods for analysing DNA, RNA and proteins share the same principle with Southern-, Northern- and Western-blot analyses, that is, they all rely on hybridization between immobilized biomolecules on a solid support and labelled mobile biomolecules in solution. In Southern-, Northern- and Western-blot analyses, the immobilized molecules are of unknown identity/abundance and are interrogated by the labelled mobile molecules of known identity. In microarray-based methods, the reverse is true. Although they share this principle, microarray analysis has a number of advantages over other techniques. The most noteworthy is the systematic approach that may provide holistic biological insights, increased throughput resulting from simultaneous analysis of a large number of genes, and potentially increased accuracy by miniaturization (as discussed in detail in *Chapter 4*).

Similarly, a number of microplate-based techniques have been developed and offer medium to high throughput in the analysis of DNA, RNA and proteins. Some of these techniques rely on hybridization but others do not. As most microplate-based techniques are easier and cheaper to set up than microarray analysis, they provide a more accessible platform and are already used in many laboratories. There are also techniques (not described in this volume) in which microarrays merge with microplates. For example, ArrayPlate (High Throughput Genomics, www.htgenomics.com) is a platform on which arrays of oligonucleotides are anchored in each well of a microplate. In Phenotype MicroArrays (Biolog, www.biolog.com), changes in different cellular properties are measured in parallel in different wells of a microplate.

Currently, there are only a few microarray books in the literature, mostly focusing on the technical aspects of hardware and on RNA analysis. This volume is intended to address a wider range of applications of microarray and microplate technologies in biomedical research. Often, there will be complementarity both where high throughput microplate techniques match low density microassay techniques; and also where microplates can permit focus on specific subsets identified by systematic microarray experiments or where microarrays represent logical scale up following hypothesis-based microplate experiments. Clearly, it is impossible for a single volume like this to cover all the different technologies in this rapidly expanding area, and we apologize to the inventors of those that are not included here. However, we hope that these pages will provide a good introduction from which the readers will then identify their own lists for further reading.

We would like to thank the contributors who devoted their valuable time to write the chapters and share their expertise with the readers. We are grateful to BIOS Scientific Publishers for the invitation to edit this book and to its staff, particularly Dr Nigel Farrar, Debora Bertasi and Eleanor Hooker for their suggestions and assistance. Mrs Diane Brown is thanked for her excellent secretarial work.

Shu Ye
Ian N. M. Day

Instrumentation for Genomics and Proteomics

1

Sarah K. Stephens, Mark R. Truesdale and Julian F. Burke

1.1 Introduction

Microplates and robotic instrumentation have become familiar tools in many laboratories, particularly those involved in high sample throughput projects, such as the Human Genome Project, and in drug discovery laboratories, which must screen thousands of lead compounds. In this chapter we cover the types of microplate currently available, their range of applications and the instrumentation that is available to handle these.

1.2 History of the microplate

According to Dr Roy Manns (1999) the history of the multi-well microplate can be traced back to the early 1950s when Dr G. Takatsky first used strips of 12 wells machined from acrylic and shortly afterward developed a 96-well version of his new plate. These early plates were machined from blocks, but by the 1960s 96-well plates were being moulded in polystyrene and other plastics, and had become commercially available. The familiar 8 × 12 cm footprint quickly became the industry standard. In the early 1990s, Genetix became the first commercial manufacturer of the 384-well plate and this format has become the standard plate in use for the majority of genomic and proteomic applications. Many high-throughput screening laboratories now use 1536-well microplates and even higher well densities have been proposed. Microplates are now available in a variety of plastics and well formats, but most adhere to the 8 × 12 cm format and many follow standards laid down by the Society for Biomolecular Screening (www.sbsonline.org). The main features of the standard are alphanumeric referencing of the well locations, 8 × 12 cm footprint, 9 mm well-to-well pitch for 96-well plates, 4.5 mm for 384-well plates and 2.25 mm for 1536-well plates.

The large number of samples or assays that can be processed in a parallel fashion when using microplates created a demand for automated means of handling the plates. The adoption of standard formats for microplates greatly facilitated this because, whereas human operators are very good at coping with samples presented in varied ways, automated systems can only work well with standardized formats.

Microarrays & Microplates: Applications in Biomedical Sciences, Shu Ye and Ian N.M. Day

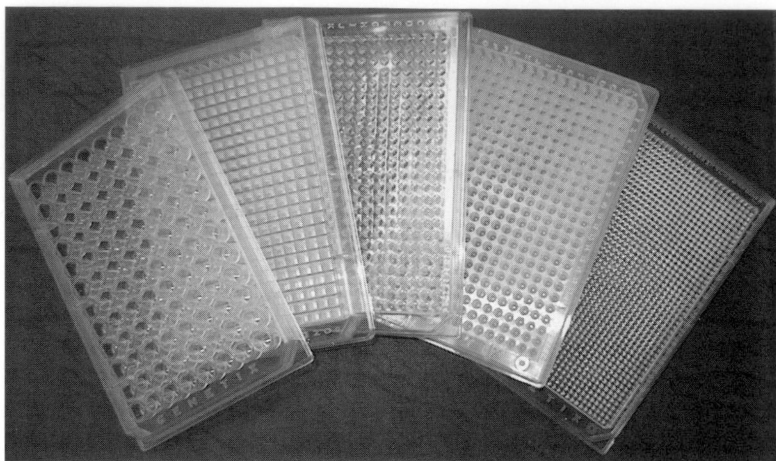

Figure 1.1

Examples of microtitre plates available in different plastics and well formats.

Microplates to suit all types of application are commercially available (*Figure 1.1*). Polystyrene plates are standard for applications such as bacterial colony picking for DNA sequencing and for assays in which the plate is 'read' by a plate spectrophotometer. Polystyrene has the advantage of being easy to mould accurately, good rigidity (excellent for handling by robotic systems) and excellent clarity for plate reading. Polypropylene is also widely used, particularly in plates for polymerase chain reactions (PCR), as it has a lower binding capacity for DNA and protein than polystyrene and can be moulded thinly to facilitate rapid heat transfer. Polypropylene plates are also commonly used in applications such as microarraying, in which DNA or protein samples are stored in the plate for long periods. As well as the plate material, the form of the wells can be chosen to suit the application, for example, square wells to maximize volume within the given area, V-bottom wells in which small volume samples must be accessed. Deep well microplates, often called 'blocks', are frequently used for larger volume applications such as growing cultures for plasmid preparations. In addition to the type of microplate that acts solely as a container for the samples, plates are available with functional coatings (e.g. streptavidin to bind biotinylated molecules), and with various types of filter incorporated in the base (e.g. for purification on the basis of molecular mass cut-off or for DNA binding and elution). It is thus possible to choose, from the huge range of plates now available, a plate that can be adapted for the automation of many processes in the molecular biology, chemistry or diagnostic laboratory.

1.3 Microplate-handling instrumentation

1.3.1 Specification of automation projects

When deciding to introduce automated systems into the laboratory it is very helpful to first specify exactly what operations you hope to perform on the new system and what degree of automation and throughput you are aiming for. Many of the problems encountered with these systems are due to the user trying to perform operations for which their robot was not specified or designed. Some systems offer a true 'walk-away' solution with the operator only having to load and unload plates at the beginning and end of a run. Others are only semi-automated; they may have 96- or 384-pin tools or pipetting tips but the user has to manually change plates. In many cases, for example, plasmid DNA preparations, a manual method of carrying out the procedure exists. In this case, it is very helpful to draw the process as a flow diagram and identify any steps which would form bottlenecks or which would be difficult to achieve on an automated system. An example is shown in *Figure 1.2*, a flow diagram for automation of a yeast two-hybrid assay. Modifications to the procedure can then be considered to avoid these

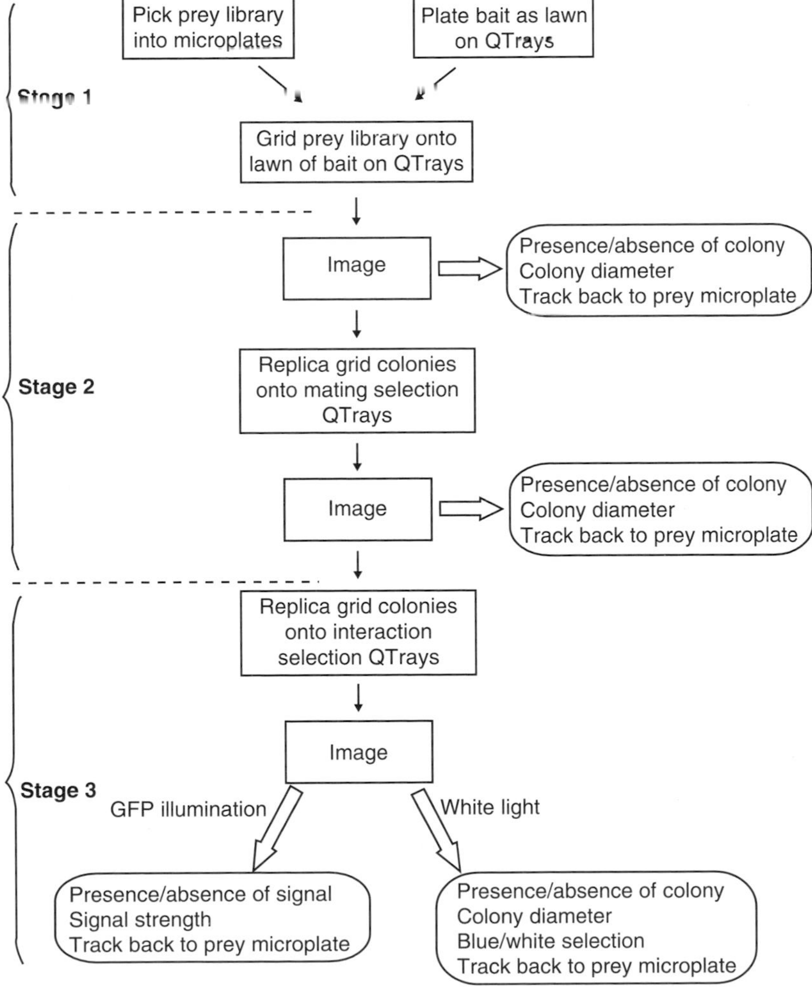

Figure 1.2

Flow diagram produced at an initial stage of specifying an automation project for yeast two-hybrid assay. Reproduced courtesy of Dr S Richmond (Genetix Ltd).

or make them more automation-friendly, or it may be accepted that manual intervention is required at some stages of the process. Some procedures, for example, high-density arraying, do not have an equivalent manual procedure and can only be achieved using appropriate robotic systems. Once an outline specification for your process is in place you can consider which of the many commercially available microplate-handling robots might fit your needs or whether for your application you need to consider commissioning a custom-built solution. Some of the common microplate-handling operations are considered below.

1.3.2 Plate handling and access

The choice and type of robotics for automation of microplate-based operations are now very wide. The specification for the process to be automated should assist in making the most appropriate choice. You need to consider how many plates the robot will need to handle at a time, whether the plates are to stay in one position or move around (e.g. in and out of a stack, or on and off an incubation station) once placed on the robot, and whether the plates are to be loaded with lids on or off. The most versatile robots for accessing and handling microplates are triple-axis gantry robots which are able to move within the X/Y and Z directions within the workspace. An example of this type is the Genetix QBot (*Figure 1.3*), this robot can access agar trays or Petri dishes on the robot bed and microplates placed either on the bed or stored in 'plate hotels'. The range of movement of the triple-axis-type robots means that they are readily adaptable to different processes simply by changing the objects supplied to the robot (pin heads, pipetting heads, microplates etc.). The QBot, for example, can be used for colony picking, gridding (macroarraying) of colonies or DNA and liquid handling processes. Other robots are available that are designed for specialist applications, for example DNA preparation, and these often have a more restricted range of movement, for example, X and Z only. This makes them less adaptable to varied processes but also tends to make them more compact. Robotic arms are also available which have limited range of movement but are designed to access microplates from a stack or carousel and feed them to a further system. For example, in a laboratory using a colorimetric diagnostic assay, a

Figure 1.3

The Genetix QBot is a floor-standing, high-throughput colony picking and macroarraying robot that has been used extensively in genome sequencing projects.

robotic arm system might be used to remove plates from a stack and place them into a microplate spectrophotometer for automated reading of the result.

The use of plate stackers, hotels or carousels increases the number of plates that can be handled in one run compared with placing plates on the robot bed. Using the robot bed only means that the robot area needs to be large if large numbers of microplates are to be loaded. The use of hotels or carousels means that the robot has to be able to pick up the microplate and move it to the area where it will be worked on. This integrates well in systems in which the plates must have lids removed or be handled on and off different workstations within the robot. A disadvantage of systems with more complex plate-handling systems is that the failure rate in a run (often caused by problems such as a plate misloaded in a stack or a plate with a cracked lid) will necessarily be higher than for a simpler system in which the microplates remain on the bed throughout. In choosing a robotic system you should consider the recovery procedure in the event of failure. Is it easy for the user to clear the problem and does the system re-set itself so that the whole run is not lost? For operations in which long run times or overnight runs are likely it is useful to consider options for remote error reporting. Systems exist whereby error messages generated by the robot can be sent to a remote PC, a pager or even a telephone.

1.3.3 Sterility

For microbiological processes sterility of the workspace and avoidance of cross-contamination between wells needs to be considered. For 'sterile' applications the workspace should be enclosed and be easy to clean between runs. Many robots are fitted with germicidal UV lights to allow disinfection of the workspace before use. For highly contamination-sensitive work, for example, tissue culture, the workspace may require a high efficiency particulate air (HEPA) filtration system. The parts of the robot that come into contact with the microbial or cell culture will need either to be disposable, for example, plastic pipette tips, or there must be a system for washing and sterilizing, for example, colony picking pins. When handling *Escherichia coli* colonies or cultures an ethanol bath is normally sufficient, but for more robust organisms, such as some spore-forming bacteria and fungi, multiple bath systems, for example, sodium hypochlorite followed by ethanol, or high temperature pin sterilization may be necessary. The user should also think about good microbiological practice when working with the robot. For example, in a case in which microplates are placed on the bed without lids they should be placed working from the back toward the front of the machine to avoid working over open microplates.

1.3.4 Imaging systems

For some applications imaging systems are required on the robot. The commonest is colony picking from agar to microplates. Here a camera is used to image the colonies. This is normally a CCD camera producing a grey-scale image; the colonies are discriminated from the background on the basis of their contrasting grey-scale values. The robot software then processes the image data to identify colonies that meet criteria set by the user such as

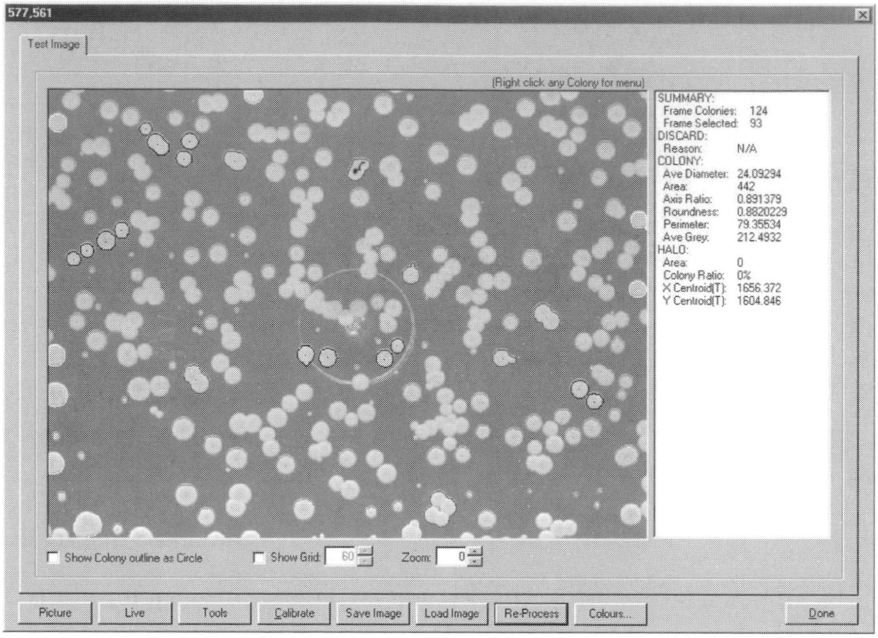

Figure 1.4

The software indicates the colonies recognized by the imaging system and marks those that meet the user's picking criteria.

minimum and maximum colony diameter, and roundness (*Figure 1.4*). In many DNA-cloning experiments the technique of blue/white selection is used in which clones lacking an inserted piece of DNA will express the enzyme beta-galactosidase and produce a blue colour on a medium containing the substrate X-Gal. These blue colonies can be discriminated from the white colonies on the basis of their grey-scale range. An option on the Genetix picking robots is the use of 'halo recognition', whereby colonies are selected on the basis of a ratio between colony diameter and a surrounding halo. This halo can be produced by antibiotic production from colonies killing a circle in a lawn of sensitive bacteria or by extracellular enzymes reacting with substrates in the medium. This function is useful in carrying out mutation screens. A recent development in colony picking technology has been the use of vectors enabling the clones to express fluorescent proteins (e.g. green fluorescent protein from *Aequoria victoria*). This requires the picking system to excite the fluorescence with the appropriate excitation wavelength, detect the emitted light and quantify the expression level of the fluorescent protein on the basis of the light emitted.

Once the desired colonies have been identified their *x/y* coordinates are used by the robot software to produce a 'script' of movements to pick the colonies and inoculate a liquid growth medium in the destination microplates (*Figure 1.5*).

Barcode reading on plates can be achieved using separate barcode scanners mounted on the robot. These are commonly used in cases in which microplates are sequentially removed from a stack or hotel to a defined bed location where the barcode scanner can be placed. Alternatively, a camera

Figure 1.5

The picking pins are fired individually into the colonies following the *x*/*y* coordinates created by the picking software.

mounted on the moveable head of the robot can be used to read barcodes on microplates at different locations on the robot bed. This camera may be the same as is used for other imaging requirements such as colony picking. The software should have options for recovery of the run in the event that a plate barcode cannot be read, because it is missing or damaged for example, perhaps giving the user options to manually enter a barcode or generate a new code for that microplate.

1.3.5 Data tracking

Although it is outside the scope of this chapter to go at length into the issue of tracking data, when working with automated systems and samples in microplates, it is an important one that should form part of the specification for the system. For some laboratories a simple manual record keeping system may be sufficient; at the other end of the scale are complex laboratory information management systems (LIMS). These systems allow all information on the samples, how they are handled and experimental results to be integrated in one database. Many of the microplate-handling robots are able to read and record barcodes on microplates and then use this information to build a log file which can be output to a LIMS. Microarraying robots, in particular, normally have a 'spot tracking' function. This takes information on the contents of each source plate well from a database, combines this with information on the microarray layout created and outputs a file with spot coordinates plus sample identities in a format compatible with the software to be used in analysis of the microarray image. The type of data tracking required from your automated system and whether this has to integrate with other systems in use in the laboratory, for example, LIMS or analysis software, is an important consideration that should not be overlooked.

1.3.6 Consumables

The consumable items (microplates, filter-bottom microplates, pipette tips, etc.) to be used on the automated system need to be considered carefully. They need to be of good quality manufacture and designed with robotic systems in mind. As mentioned earlier, although a human operator can cope with variations, an automated system will not do so very easily. The high throughput that can be achieved using robotics means that use of consumables will increase, so cost is an important consideration as is storage space, such as a −80°C freezer, for the microplates once generated.

1.3.7 Limitations of automated instrumentation

It is a common misconception that an automated system will always carry out a given process better than a skilled manual operator. This is, of course, the ideal that is aimed at when designing a system, but it is not necessarily true in all cases. Automated systems avoid simple human errors such as placing the pin or pipette tip in the wrong well but they are not so good as a human at coping with variation in a system. For example, in colony picking a human operator can easily compensate for variations due to agar plates poured to different depths or not set level, and even variations in the blue/white discrimination of colonies due perhaps to plates having different quantities of X-Gal or being incubated for different times. An automated vision system and robot will not cope well with variations of this nature, nor will it cope well with being fed plates with warped or cracked lids, which a human operator would have no problem with. Thus, when commissioning an automated system in the laboratory these factors external to the robot merit as much attention as the robot itself. It is a great help to have standard operating procedures to standardize and control the preparation of materials that will be fed to the automated system and this will reduce the number of problems encountered. An example in use at Genetix is shown in *Protocol 1.1*.

However, in some cases it has to be accepted that the percentage failure rate (e.g. in colonies that fail to be picked) is higher with the automated system than the manual, but this is compensated for by the much higher throughput. A skilled technician can pick several hundred bacterial colonies per hour, and will be able to make flexible decisions on which colonies to pick and where on the colony to pick them. The typical example being choosing to pick two colonies that are in very close proximity, but then pick at the furthest edge of each to minimize the chance of cross-contamination. A robotic system can pick around 4000 colonies per hour, but it will only pick colonies meeting the criteria set up by the user at the start of the routine. It will not make the type of 'flexible decisions' that a human would, thus a higher percentage of colonies on a plate will 'fail', for example, because two colonies are in close proximity. These factors need to be considered when specifying and choosing automated systems for the laboratory.

1.4 Applications in genomics

1.4.1 Colony picking

The high-throughput genome sequencing projects (International Human Genome Consortium, 2001; Venter *et al.*, 2001) that have been so successful in recent years have relied on the development of automation based around standardization of sample handling in microplate format (Jaklevic *et al.*, 1999). One of the key stages in any genome sequencing project is the picking of clone libraries prior to the preparation of the DNA sequencing template. A skilled human operator can pick up to 600 colonies per hour. Some early designs of picking robot actually picked at a slower rate than this (Uber *et al.*, 1991), but current models, such as the Genetix MegaPix and QPix, will pick ≈3500 colonies per hour. A further advantage of producing libraries arrayed in well plates is that they can then be stored as a permanent and indexed resource, copied and used for other purposes (Jones *et al.*, 1992). The production of gene libraries arrayed in microplates has now been widely applied to many different organisms for the purpose of genome sequencing and for screening libraries for particular genes and for gene expression analysis, for example *Arabidopsis* (Giege *et al.*, 1998) and sea urchin (Cameron *et al.*, 2000).

1.4.1.1 Practical considerations for colony picking

The most important points to ensure successful colony picking are making sure that the amount of agar poured is consistent and that the Petri dishes or trays are set level. The destination microplates must be filled with a consistent volume of growth medium. The labour involved in filling these microplates is greatly reduced by using an automated plate filler such as the QFill (*Figure 1.6*). The number of colonies spread on the agar should be sufficiently dense to avoid wasting materials but not so dense that the colonies must be picked when they are very small or that many colonies fail to be picked because of their close proximity to adjacent colonies. We generally aim for ≈3500 colonies on 22 × 22 cm trays. For spreading colonies on large trays we use sterile glass beads of ≈5–6 mm diameter and roll these around the plate to spread the colonies. For the larger trays this is not only quicker than using a traditional glass spreader, it also results in a more even spread of colonies.

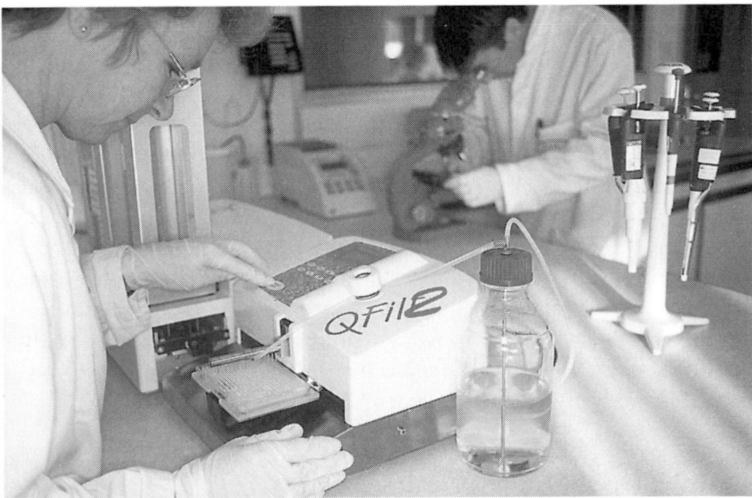

Figure 1.6

An automated well plate filler, such as the QFill, is a valuable addition to any laboratory using microplate-based procedures.

In the robot software it is important to adjust parameters such as colony diameter, roundness, etc. to take account of the colony size and morphology of the organism that you are picking. The grey-scale threshold for colony recognition may also need to be adjusted for colonies of different colour and opacity. *E. coli* colonies are generally straightforward to pick as most colony-picking robots were designed with these in mind. However, modifications such as using different pins or sterilization procedures mean that the robot can be adapted to pick different organisms.

For example, for *Saccharomyces cerevisiae, Schizosaccharomyces pombe* and other yeasts, larger diameter yeast picking pins may be useful, although many yeasts can be picked with standard picking pins, which are 0.6 mm diameter. However, some yeast strains are susceptible to drying out on the pin between picking and inoculation of the well plate, thus reducing picking efficiency. In this case, the problem can normally be resolved by using a grooved picking pin. A single 80% ethanol wash bath may not be sufficient to sterilize some yeast species. If this proves to be the case use a three-bath system with: bath 1, 1% sodium hypochlorite; bath 2, water; bath 3, 80% ethanol. If contamination is still proving a problem make the second wash bath 80% ethanol also. This solution has prevented carry over with a number of yeasts tested in our laboratories.

Our experience with *Streptomyces* and other filamentous organisms is that pins with a barbed tip design ('Christmas tree pins') are optimal for picking these organisms. To ensure sterilization of the pins use the three-bath wash system described above or a high temperature pin dryer. These organisms often form colonies that are not perfectly circular (and have ragged edges) so you will almost certainly have to reduce the stringency of the software picking parameters for colony roundness in order to select the colonies.

Some organisms may not grow well if picked into liquid culture. In this case colonies can be picked onto agar in 8×12 cm trays or in the wells of microplates. You will need to adjust the inoculation depth to less than used when inoculating liquid media so that the pins do not stab too deeply into the agar.

Bacteriophage M13 plaques are produced in the standard way by spreading an agar or agarose layer containing the phage suspension onto an agar medium. For phage producing small, slow growing plaques, for example, M13, it may be necessary to incubate the plates longer than normal in order for the plaques to be large enough for the robot imaging system to recognize.

Bacteriophage λ plaques are normally fairly straightforward to image. The picking pins used are generally larger in diameter than those used for *E. coli* colonies, we normally recommend 1.6 mm pins. Rather than picking into a sterile growth medium, the phage plaques can be picked into microplates containing cultures of a host strain in order to produce a small-scale phage lysate. Alternatively, the phage plaques can be picked directly into a phage buffer (e.g. SM buffer with 7% dimethylsulfoxide [DMSO]). Enough phage DNA will be present to allow amplification by standard PCR methods (R Davies, personal communication).

1.4.2 Macroarraying (gridding)

The quickest and least costly method of producing DNA arrays from a clone library arrayed in microplates is to use a robotic arrayer or gridder to produce high-density colony arrays on nylon membrane. Unlike glass slide microarrays, the cost of producing the array is sufficiently low that it is not necessary to fully characterize the clones and eliminate redundancies before arraying. The most common formats for these arrays are 8 × 12 cm and 22 × 22 cm. Arraying is achieved using 96- or 384-pin tools to transfer bacterial cells from the microplate array to the membrane surface. An offset pattern is built up to increase the density of the array produced. Bacterial colonies are grown on the membrane surface and then treated to lyse the cells and bind the released DNA to the membrane. Hybridization is then carried out using standard techniques with radioactive (typically ^{33}P) or non-radioactive methods, such as digoxigenin (DIG) (Roche, Lewes, UK) or enhanced chemiluminescence (ECL; Amersham Pharmacia Biotech, Amersham, UK). Imaging of the results can be by autoradiograph, phosphorimager or CCD camera system as appropriate to the detection method.

The arrays have a wide range of applications from simple library screening for a particular sequence to more complex applications such as gene expression analysis. Cox (2001) gives a review of the use of nylon arrays for expression analysis. The detection limit quoted for radioactive detection is 1 transcript in 10^6 when starting with 1 μg of mRNA, and this limits the utility of the method described for analysis of rare transcripts. A further consideration is that rare transcripts may not be represented in the arrayed cDNA library in the first place. These issues are addressed by the method published by Rast *et al.* (2000). Their method uses large cDNA libraries and a method of subtractive hybridization between 'tester' and 'driver' populations of mRNA with limited amplification steps to enrich for differentially expressed rare transcripts. The group works with sea urchin embryos; an example of their application is given in Ransick *et al.* (2002) where genes involved in endomesoderm specification early in embryonic development were identified. However, the method is generally applicable and particularly useful for organisms in which gene sequence is limited making microarrays impracticable. In our laboratories we have used the method to compare *Drosophila* embryos with adults and males with females. The necessary reagents for the subtractive hybridization and the amplification primers are available as the Genetix GeneSeeker kit.

A laboratory that has developed a range of array applications is that of Professor Hans Lehrach (Clark *et al.*, 1999). As well as gene expression work, they have developed other applications, for example, sequence fingerprinting, using a series of automated systems to process libraries arrayed in microplates, high-throughput PCR of the inserts, arraying of the DNA and hybridization to octanucleotides. The method is used to group unknown clones into those containing the same or closely related genes (Maier *et al.*, 1994).

1.4.2.1 Practical considerations for gridding

Gridding of bacterial colonies from microplates onto membranes can be done by arraying directly onto the membrane laid on agar in 22 × 22 cm trays or onto membrane laid on the robot bed. Gridding robots are supplied with special blocks to provide a flat surface for gridding. We use 96- or 384-pin gridding heads with 0.4 mm gridding pins for colony arrays. Using these pins we array 55 296 colonies onto a 22 × 22 cm membrane. The colonies are arrayed in duplicate so this represents 27 648 individual clones, which is 72 × 384-well source microplates (*Figure 1.7*).

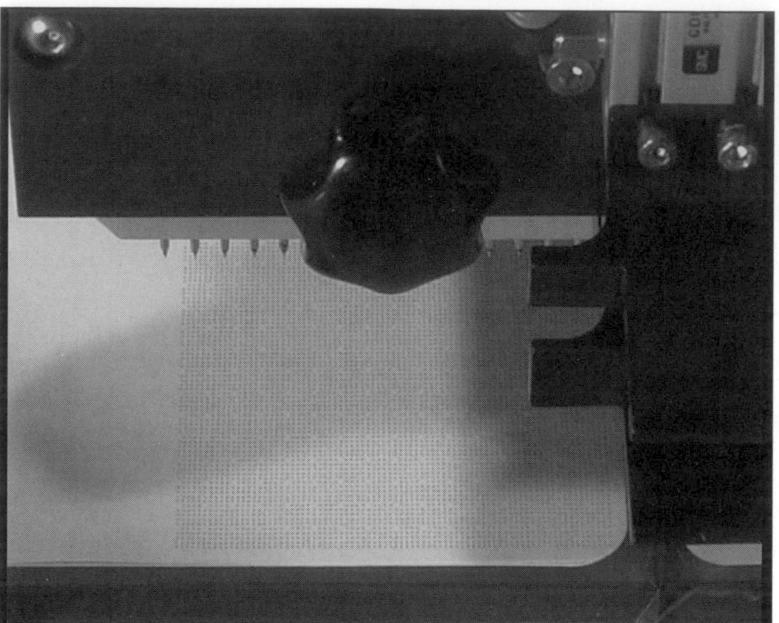

Figure 1.7

Macroarraying is achieved using 96- or 384-pin arraying heads to transfer sample from microplates to membrane.

If arraying directly onto agar a gridding head with gravity pins must be used as a sprung head tends to pierce through to the agar. If arraying on membrane laid on solid blocks soak a piece of 1.5 mm blotting paper in growth medium and lay it on the block. Roll with a sterile 10 ml pipette to remove air bubbles and excess liquid. Soak the membrane in the medium and lay on top of the blotting paper and roll again. This can be done working in a laminar flow hood placed near the robot to assist in keeping the materials free of contamination. Use a nylon membrane, for example, PerForma (Genetix), rather than nitrocellulose. Nitrocellulose is more difficult to handle and tends to tear easily, especially when used in sheets of 22×22 cm. PerForma membrane is cast onto a more rigid substrate than standard nylon membrane, which makes it easier, and therefore quicker, to handle when setting up the robot for gridding. After gridding the membranes are incubated in order to allow colony growth on the surface. If the colonies have been gridded directly onto membrane laid on agar simply replace the plate lids and invert the plates for incubation. If you have gridded onto membranes laid on blocks then the membrane must be carefully lifted by the corners and placed onto agar before incubation.

With some *E. coli* strains (notably SOLR and some BAC libraries) it may be found that growth after gridding from glycerol stocks is poor. Some gridding robots give the option of stirring the source plates or making multiple transfers of sample to the same spot, both of which can assist in transferring more colonies to achieve satisfactory growth. If this does not work the plates will need to be replicated to fresh medium without glycerol and gridding is then carried out from the freshly grown plates.

After growth of the colonies the membrane is treated to lyse the cells and bind the DNA to the membrane. The protocol that Genetix recommend includes a proteinase K treatment step which helps to remove cell debris from the membrane and reduces background in subsequent hybridization (*Protocol 1.2*).

If you are spotting PCR products or other DNA it is not necessary to purify the PCRs first. There are two commonly used protocols:

1. Place a sheet of blotting paper moistened with a denaturing solution (e.g. NaOH) onto the gridding block and then place the nylon membrane on top, so that you are arraying onto a slightly damp membrane. The DNA is denatured as it is printed. After printing neutralize and UV cross-link/bake the membrane as you would for a normal dot blot.
2. Add a denaturing solution to the DNA in the microplate and then grid onto a dry membrane.

The second protocol has the advantage that smaller spots are produced, enabling higher density arrays, but DNA that has been suspended in a denaturing solution may not be suitable for storage and re-use of any residual sample.

1.4.3 DNA preparation

The onset of high-throughput genomics also brought about the need for high-throughput DNA purification systems that could produce high-purity DNA quickly and efficiently. The use of microplates is again central to this. More than 10 years ago when the first plasmid miniprep and PCR clean-up kits appeared they used individual spin column formats. These same technologies have been scaled up to microplate format.

For microplate-based DNA purification there are a few basic alternative technologies. The first, and most commonly used, is based on the ability of glass fibre membrane to bind DNA in the presence of a high-chaotrophic salt such as guanidine hydrochloride. This basic principle allows the selective removal of DNA from an aqueous solution such as a PCR reaction or bacterial lysate (for plasmid DNA). The membrane can then be successively washed to remove any impurities and the pure DNA is then eluted from the membrane in a low salt buffer. There are now several commercial kits available which use this technology in a 96-well format, and many of these have been used by instrument manufacturers as the basis of automated DNA purification systems, for example, Tecan, Qiagen and Packard.

An alternative to this technique is paramagnetic particle (PMP) technology, such as seen in Promega's MagneSil™ kits. This technique utilizes selective binding of DNA to specific type PMPs. The PMPs are mixed with the impure sample, the DNA binds to the PMPs which can then be attracted to the sides of the microplate well using magnetic pins, allowing the supernatant to be removed. This technology removes the need for centrifugation or vacuum steps, which are typically required by the earlier technique.

At Genetix, we have recently developed our own microplate-based PCR and plasmid DNA systems, called genPURE. This system is based on the glass fibre membrane principle. This procedure is easily automatable, and we have developed a Genetix QBot for this process. With this product we are able to produce a good yield of high-purity DNA from in excess of 7000 plasmid samples in a 24-h period.

1.4.4 DNA microarrays

The term DNA microarray is generally used to cover arrays of DNA spotted at high density on solid surfaces, normally a treated glass slide. Detailed discussion of microarray theory and techniques will not be given here as it is covered in other chapters. Applications for which DNA microarrays are being applied include gene expression, SNP genotyping, gene dosage studies, toxicology and sequencing by hybridization. A Nature Genetics Microarray supplement published in 1999 gives a good introduction to the various application areas. There are now many excellent web resources published by leading microarray laboratories such as the Pat Brown laboratory at Stanford University, the Ontario Cancer Institute and the Institute for Genome Research. This latter group have published an excellent review of their techniques, which is a good starting point for anyone new to microarraying (Hegde *et al.*, 2000).

1.4.4.1 Practical considerations for DNA microarrays

The DNA to be arrayed is containing in microplates, the most common format being 384-well. Many laboratories prefer polypropylene microarray plates for their higher hydrophobicity and lower DNA binding capacity. The DNA to be arrayed is most commonly PCR products derived from known sequence clones or synthetic oligonucleotides. The spotting buffer in which the samples are dissolved should be compatible with the slide surface chosen, optimized spotting buffers are now commercially available. There are two common classes of microarray slide chemistries:

1. Those relying on a non-covalent surface interaction, for example, e.g. poly L-lysine, silane or amino propyl silane (e.g. Corning CMT-GAPS, ArrayIT™SuperAmine); and
2. Those using a covalent bond formed via Schiff's base chemistry (Silylated slides, ArrayIT™ SuperAldehyde).

This latter type relies on the interaction between an amine group on the DNA and an aldehyde group on the substrate surface. Typically a 5′-amino group can be added to the DNA by using 5′-amino-linker PCR primers to amplify the DNA to be arrayed. Oligonucleotide arrays can be printed in the same way as cDNA arrays but for best results we would recommend using oligos with a 5′-amino group and using a covalent attachment slide chemistry.

There are now almost a dozen different companies producing microarraying robots. The principle is similar to that of gridding, except that a smaller number of pins are used to produce densely packed arrays that fit the typical microscope slide format (*Figure 1.8*). Also a higher degree of resolution, particularly on the X and Y drives is demanded to enable accuracy when arraying at high density. There are basically two types of arraying robot: those that use pin or contact printing and those that use piezoelectric inkjet printing. The former is the most commonly used at present. The printing pins may either be solid tip pins or one of several different designs that rely on capillary action to load a defined volume into the pin. The solid pins must be dipped into the sample between each spot, whereas the other types are multi-spotting and a defined volume is dispensed each time the pin contacts a slide. This avoids the need for repeated reloading of the pins and makes printing of large numbers of arrays much quicker. Solid pins

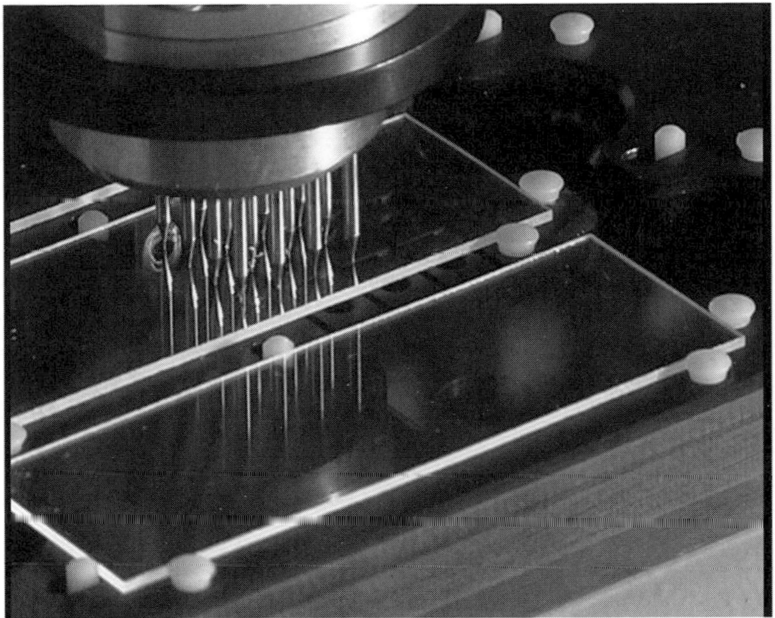

Figure 1.8

Microarray robots transfer DNA samples from source microplates onto glass slides treated with a coating to enable DNA binding.

have the advantage of being cheaper, more robust and not prone to blockage. The pins are washed and dried in between pick up of different samples to avoid carry over between spots. Washing is normally achieved with water to remove DNA from the pins and ethanol to promote rapid drying. Drying is accomplished by high-velocity airflow produced either via a vacuum pump or compressor.

In choosing an arraying robot the user should consider the number of arrays that they are likely to produce and the spot density of those arrays. Arraying robots vary widely in capacity for slides and source plates and unattended run times, and the price of the machines varies accordingly. Consideration should also be given to requirements for environmental controls, such as HEPA filtration and humidifier; these are standard on some machines and optional extras on others. They may or may not be required depending on the laboratory conditions under which you intend to site the machine and the run times that will be required. Generally the most satisfactory and reproducible results will be obtained if the machine can be sited in a room with air conditioning that can be kept relatively clean and dust-free.

In addition to the purchase of the arraying robot, a microarraying laboratory will require a slide scanner. There are two types of optical system available, either confocal laser or CCD camera. CCD systems have the advantage of being generally faster but confocal laser systems are generally regarded as being more sensitive and having higher resolution. The recently released Genetix aQuire scanner has confocal laser optics. All microarray scanners on the market are configured to detect the most commonly used fluorescent dyes for microarray analysis, which are Cy3™ and Cy5™. However, many are also equipped with additional lasers and filters to enable a wider range of dyes. High-throughput laboratories may also want to consider scanners which have options for multiple slide loading systems otherwise scanning large numbers of slides will be labour intensive.

1.5 Applications in proteomics

The success of the high-throughput approaches to genomics has led to the birth of proteomics in which the total protein content of the cell, the 'proteome', will be studied in a similar high-throughput and parallel fashion. Again the microplate array will be the basis for automation of assays and sample handling and many of the systems developed for genomics applications, such as arraying, are also applicable to proteomic applications. There are, however, special considerations when handling proteins as opposed to bacterial colonies or DNA. Proteins are much more diverse in composition and, as such, different samples may require different preparation and handling methods. In general proteins are much less robust than nucleic acids and may require more stringent environmental control such as temperature and humidity than is necessary for DNA arraying. The inherent variability in viscosity between different protein samples means that arrayers with solid pins are most commonly used to produce the arrays. The multi-spotting pins commonly used for DNA arrays do not deliver reproducible amounts where sample viscosity varies widely and, indeed, may fail to load at all where viscosity is very high. At present, most protein arrays are being produced on poly(vinylidene difluoride) (PVDF) membranes or on glass slides designed for DNA arrays but it seems certain that surfaces specifically optimized for protein arrays will be developed and commercially available in the near future.

1.5.1 2D gel picking and mass spectrometry

2D gel electrophoresis is an old technique for protein expression profiling and, despite its drawbacks in terms of ease of use and reproducibility, it has yet to be surpassed for sensitivity and resolving power for complex protein samples (Jenkins & Pennington, 2001). In the new proteomics applications scientists are seeking to analyse large numbers of the protein spots from gels. This has necessitated the development of automated spot imaging and excision systems that can identify and pick spots from gels into arrays in microplates. Once in the microplate the typical workflow is to carry out an in-gel digestion of the protein and elution of the peptide fragments which are then analysed by mass spectrometry. Matrix-assisted laser desorption ionization (MALDI) mass spectrometry requires the peptides to be placed on a solid matrix for analysis and this can be achieved using arraying robots in a similar fashion to produce macroarrays as described previously.

The rate of analysis of a mass spectrometer is now very high so that the bottleneck in the process is likely to be in the gel spot picking or in the digestion step. Laboratory automation companies are thus concentrating on these areas. A number of machines have recently been launched, for example, Genetix gelPix, and this area is certain to see further developments in the future (*Figure 1.9*).

Figure 1.9

The gelPix uses an eight-channel excision head to take cores from two-dimensional protein gels and transfers them to microplates for further analysis.

1.5.2 Protein arrays

The use of protein arrays is a rapidly growing field generating much interest (Cahill, 2001; Walter *et al.*, 2000), although the literature with successful examples is still rather limited. An attractive target for protein arrays are antibodies either on the array themselves or as a means of screening an array of unknown proteins. For example, Walter *et al.* (2001) created arrays of phage display proteins on PVDF filters and screened these for binding events using scFv antibody fragments. The binding events were confirmed using assays in microplate format. The whole process was highly automated using picking and arraying robots to create the arrays and automation of microplate assays. Bussow *et al.* (1998) and Holt *et al.* (2000) used a similar technique to produce arrays from fetal brain expression libraries, which were screened with antibodies to his tags plus some specific genes and scFv antibody fragments respectively.

The most useful protein arrays will be those in which the proteins are immobilized on the array in such a way as to retain their structure and function enabling interactions with other proteins, nucleic acids or small molecules to be assayed in an array format. MacBeith and Schreiber (2000) have demonstrated, with a limited number of proteins, that it is possible to produce a glass slide microarray of proteins and assay for protein functions such as protein–protein interaction. These arrays were produced using a pin-type arrayer originally designed for DNA microarrays. The slides used were of the aldehyde-derivatized type, more usually used to attach to amine-modified DNA. In this case, the aldehyde groups react with naturally occurring amine groups on the proteins. The first true proteome microarray was produced by Zhu *et al.* (2001). In this they describe the cloning, expression and arraying of proteins from 5800 yeast open reading frames. The proteins were expressed with a HisX6 tag and immobilized by interaction with a nickel-coated slide. The arrays were used successfully to probe for protein–protein interactions and protein–lipid interactions.

1.6 Conclusion

The tremendous advances in laboratory automation and the standardization of sample handling in microplate format were driven by, and essential to, the high-throughput approaches taken to accomplish the recent genome projects. It is certain that automation will play an equally vital role in the rapidly growing field of proteomics and that as new methods and assays are developed there will be a continuing demand for new developments in automation. Although the 96- and 384-well microplate looks set to remain the standard platform for the near future there is an increasing trend to assay in smaller volumes using 1536 or higher density plates and moves to microfluidic systems.

References

Bussow K *et al.* (1998) A method for global protein expression and antibody screening on high-density filters of an arrayed cDNA library. *Nucleic Acids Res* **26** (21): 5007–5008.

Cahill DJ (2001) Protein and antibody arrays and their medical applications. *J Immunol Methods* **250** (1–2): 81–91.

Cameron RA *et al.* (2000) A sea urchin genome project: Sequence scan, virtual map and additional resources. *Proc Natl Acad Sci USA* **97** (17): 9514–9518.

Clark MD *et al.* (1999) Construction and analysis of arrayed cDNA libraries. In SM Weissman, editor. *Methods in Enzymology*, Vol. 303. San Diego: Academic Press.

Cox JM (2001) Applications of nylon membrane arrays to gene expression analysis. *J Immunol Methods* **250**: 3–13.

Giege P *et al.* (1998) An ordered *Arabidopsis thaliana* mitochondrial cDNA library on high-density filters allows rapid systematic analysis of plant gene expression: a pilot study. *Plant J* **15** (5): 721–726.

Hegde P *et al.* (2000) A concise guide to cDNA microarray analysis. *BioTechniques* **29** (3): 548–562.

Holt LJ *et al.* (2000) By-passing selection: direct screening for antibody–antigen interactions using protein arrays. *Nucleic Acids Res* **28** (15): e72.

International Human Genome Sequencing Consortium (2001) Initial sequencing and analysis of the human genome. *Nature* **409**: 860–921.

Jaklevic JM *et al.* (1999) Instrumentation for the Genome Project. *Annu Rev Biomed Eng* **1**: 649–678.

Jenkins RE, Pennington SR (2001) Arrays for protein expression profiling: towards a viable alternative to two-dimensional-gel electrophoresis? *Proteomics* **1**: 13–29.

Jones P *et al.* (1992) Integration of image analysis and robotics into a fully automated colony picking and plate handling system. *Nucleic Acids Res* **20** (17): 4599–4606.

MacBeath G, Schreiber SL (2000) Printing proteins as microarrays for high-throughput function determination. *Science* **289**: 1760–1763.

Maier E *et al.* (1994) Application of robotic technology to automated sequence fingerprint analysis by oligonucleotide hybridisation. *J Biotechnol* **35** (2–3): 191–203.

Manns R (1999) Microplate history. A presentation given at MipTec-ICAR, 1999. Available at http://www.messebasel.ch/miptec/mission_history.htm Nature Genetics Microarray Supplement (1999). *Nature* **21**.

Ransick A *et al.* (2002) New early zygotic regulators of endomesoderm specification in sea urchin embryos discovered by differential array hybridization. *Dev Biol* (in press).

Rast JP *et al.* (2000) Recovery of developmentally defined gene sets from high-density cDNA microarrays. *Dev Biol* **228**: 270–286.

Uber DC *et al.* (1991) Application of robotics and image processing to automated colony picking and arraying. *BioTechniques* **11** (5): 642–647.

Venter JC *et al.* (2001) The sequence of the human genome. *Science* **291** (5507): 1304–1351.

Walter G *et al.* (2000) Protein arrays for gene expression and molecular interaction screening. *Curr Opin Microbiol* **3** (3): 298–302.

Walter G *et al.* (2001) High-throughput screening of surface displayed gene products. *Comb Chem High Throughput Screen* **4** (2): 193–205.

Zhu H *et al.* (2001) Global analysis of protein activities using proteome chips. *Science* **293**: 2101–2105.

Protocol 1.1: Arraying of PCR products onto nylon membranes

REAGENTS

1. Denaturing Solution (see Standard Method 34).
2. Neutralizing Buffer (see Standard Method 35).

METHOD

1. Fill a Bio-assay 'Q' tray with denaturing solution and briefly soak a piece of gel blotting paper, using forceps to hold the filter paper at opposite corners for ease of handling. Remove filter paper and quickly drain off the excess denaturing solution.
2. Carefully lay the filter paper onto a filter block.
3. Briefly soak a nylon membrane in denaturing solution, using forceps for handling. Remove membrane and quickly drain off the excess denaturing solution.
4. Carefully overlay on top of the filter paper on the filter block. Gently roll a sterile 10 ml disposable pipette over the surface of the membrane in order to remove any air bubbles.
5. Place the filter block on the bed of the sterilized QBot or QPix and repeat the process for the required number of membranes.
6. Gridding is now ready to commence.
7. After gridding, remove the filter blocks from the bed.
8. Fill a Bio-assay 'Q' tray with denaturing solution and using forceps, briefly soak a piece of gel blotting paper. Remove the filter paper and quickly drain off any excess denaturing solution.
9. Place the filter paper in the lid of a bioassay 'Q' tray.
10. Using forceps, remove the nylon membrane from the filter block and place it on the pre-wetted filter paper in the lid of the bioassay 'Q' tray.
11. Let stand for 10 min, then place the nylon membrane onto a second sheet of filter paper, pre-wetted with neutralizing buffer, in the lid of a bioassay 'Q' tray.
12. Let stand for 5 min and then transfer the membrane to a dry piece of filter paper and leave to air dry (over night).
13. Place the membrane in a UV cross-linker, DNA side up and expose at 120 000 mJ cm^{-2}.

Protocol 1.2: Processing colony arrays

METHOD

1. Remove the gridded plates from the incubator and transfer them to a fridge until ready to process.
2. Pour ≈200 ml of denaturing solution into a 22 × 22 cm tray and wet a sheet of gel blotting paper in the solution.
3. Remove the gel blotting paper, drain excess liquid and place the sheet on the lid of a 22 × 22 cm tray.
4. Using forceps, remove the gridded membrane from the agar plate and place it squarely onto the pre-wetted gel blotting paper for 4 minutes.
5. During the 4 minutes, wet another sheet of gel blotting paper with denaturing solution, drain excess liquid and place it on a second lid.
6. Using forceps, transfer the membrane to the second pre-wetted gel blotting paper.
7. Immediately transfer the membrane together with the gel blotting paper onto a glass plate suspended above a boiling water bath for 4 minutes.
8. During the 4 minutes fill a 22 × 22 cm tray with ≈200 ml of neutralizing solution and soak a sheet of gel blotting paper in the solution.
9. Remove the gel blotting paper, drain excess liquid and place on a tray lid.
10. Remove just the membrane and place squarely on the pre-wetted gel blotting paper with neutralizing solution. Let stand for 4 minutes.
11. Transfer the membrane to a dry sheet of gel blotting paper and let stand for 1 minute.
12. During the 1 minute fill a 22 × 22 cm tray with 100 ml of proteinase K solution.
13. Transfer the gridded membrane, colony side down, into the proteinase K solution.
14. Incubate for 1 hour at 37°C.
15. Transfer the membrane, colony side up, onto a sheet of dry gel blotting paper.
16. Place another sheet of dry gel blotting paper over the membrane and roll with a sterile 10 ml glass pipette.

17. Place the membrane between two sheets of dry gel blotting paper and dry overnight at room temperature.
18. Next day, remove the top sheet of gel blotting paper and place the membrane into an ultraviolet cross-linker, DNA side up. Expose the membrane at 120 000 μJ cm^{-2}.

Solutions

Denaturing solution:

NaCl = 87.67 g
NaOH = 20 g
dH_2O = make up to 1 litre

Neutralizing solution:

NaCl = 87.67 g
Tris = 121 g
dH_2O = make up to 1 litre; adjust to pH 7

Proteinase K:

5 M NaCl = 20 ml
1 M Tris–Cl pH 8.0 = 50 ml
0.5 M EDTA pH 8.0 = 100 ml
10% Sarkosyl = 100 ml (*N*-lauryl-sarcosine)
dH_2O = 730 ml
Incubate overnight at 37°C
Next day add 100 mg proteinase K

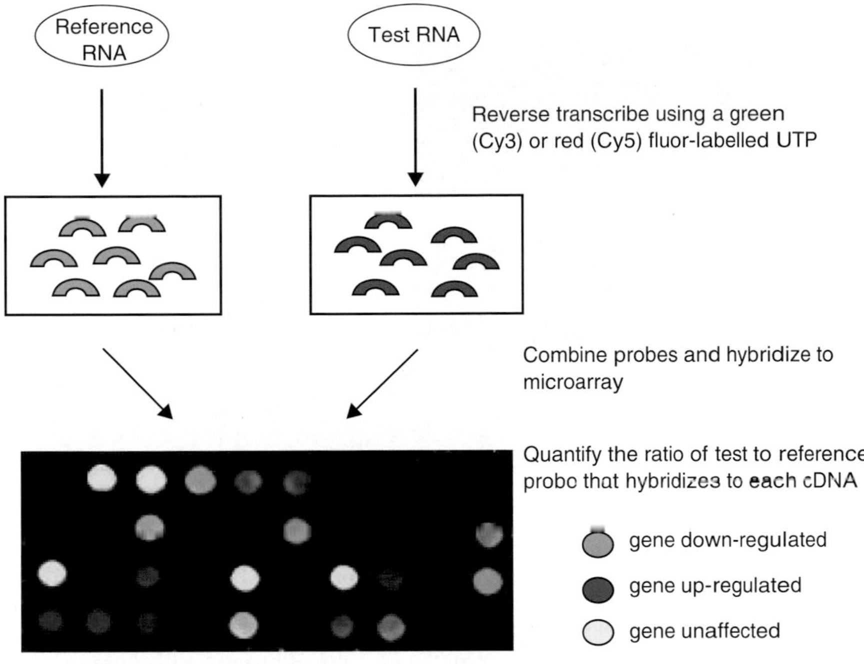

Plate 1

Preparation of fluorescently labelled probes and dual-hybridization to glass microarrays.

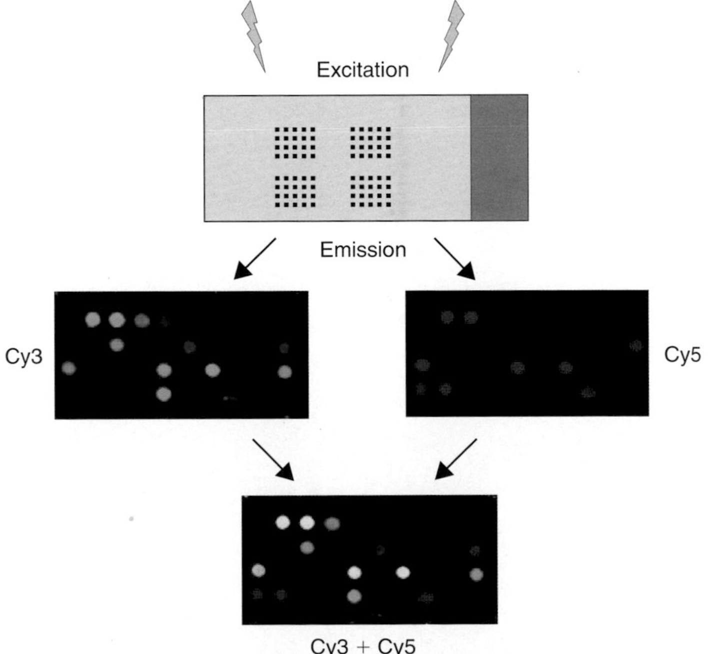

Plate 2

Scanning and image processing. Microarrays are excited with lasers of different wavelengths leading to distinct emissions from the Cy3- and Cy5-labelled probe that is bound to the arrayed DNA. Following detection, the separate image files are exported to analysis software where they are converted to pseudo-coloured images (Cy3, green; Cy5, red). A visual interpretation of expression changes is provided by merging the two images. Green spots, genes whose transcripts are over-represented in the Cy3-labelled RNA; red spots, genes whose transcripts are over-represented in the Cy5-labelled RNA; yellow spots, genes equally represented in both RNAs.

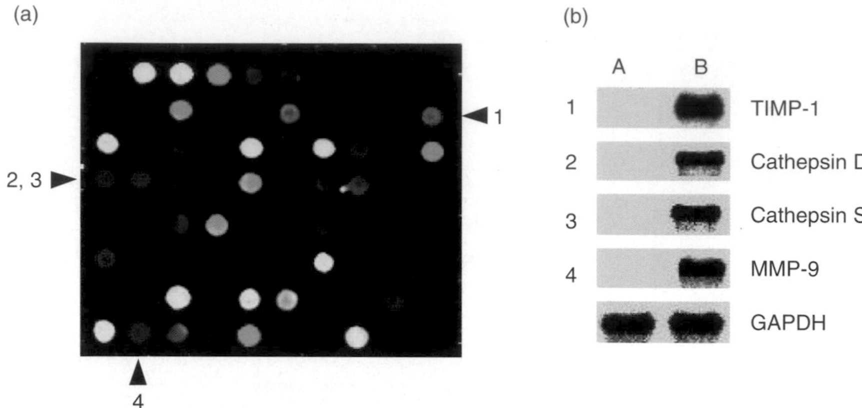

Plate 3

Microarray and Northern analysis to monitor gene expression in a human monocytic cell line before and after differentiation into macrophages. (a) Microarray experiment. mRNA from monocytic cells was labelled with Cy3-UTP (green fluor) and mRNA from macrophages was labelled with Cy5-UTP (red fluor) and the combined probes were hybridized to a microarray. Following scanning, images were processed and merged using ScanAlyze software (M. Eisen, ©1998–9, Stanford University) and saved as a jpeg file. The indicated spots correspond to: 1, TIMP-1; 2, cathepsin D; 3, cathepsin S; 4, MMP-9. All four genes are upregulated when monocytes differentiate into macrophages. (b) Northern blot. Samples of the same mRNAs were probed with cDNAs for the same four genes. Lane A, monocytic cells; lane B, macrophages. GAPDH was used as a loading control.

Expression Microarrays

2

Carl Whatling, Shu Ye and Per Eriksson

2.1 Introduction

Rapid progress in the field of functional genomics has made it possible to monitor the expression of thousands of genes simultaneously. Not only does this have the advantage of high-throughput analysis, but also the increase in the number of data points has a direct effect on the accuracy and relevance of the information generated. A variety of technological developments have made this possible, including expressed sequence tag (EST) analysis (Adams *et al.*, 1991; Okubo *et al.*, 1992), differential display (Liang & Pardee, 1992), serial analysis of gene expression (SAGE; Velculescu *et al.*, 1995) and a variety of array-based technologies (Schena *et al.*, 1995, 1996, Lockhart, 1996). Of the various options, much progress has been made in the development and refinement of array technologies, reflected by the abundance of diverse publications in which this technology has been employed (e.g. Chabas *et al.*, 2001; Hedge *et al.*, 2001; Hoffmann *et al.*, 2001, Kato-Maeda *et al.*, 2001; Thimm *et al.*, 2001). The purpose of this chapter is to provide an overview of the principles and techniques of expression arrays, with particular emphasis on the use of cDNA expression microarrays.

2.2 Microarrays in overview

The term array refers to the regular arrangement of oligonucleotide or cDNA representations of genes on a solid support such as a nylon membrane, plastic or glass microscope slide. In the latter case, it is possible to array samples very close together such that the term 'microarray' can be applied. Gene expression is monitored following hybridization with a probe generated from RNA in which each expressed gene should be represented in relative proportion to its transcript level. In principle, it should be possible to measure absolute levels of expression for each represented gene by quantifying the amount of a single probe that hybridizes to the arrayed DNAs. However, a variety of factors entail that hybridization of probe to target is not equivalent for every expressed transcript. This includes consideration of the target sequences, the labelling method and the hybridization conditions (Eisen & Brown, 1999). For such reasons, it has become standard to quantify expression levels by measuring the relative level of hybridization between two different probes. By convention, probe is prepared from two related RNA pools corresponding to a reference and test condition, and the measurement made is the amount of test probe relative to reference probe that binds to each arrayed DNA.

Probes are prepared from RNA pools by reverse transcription to generate first-strand cDNAs. Two methods have been used in particular to label

Microarrays & Microplates: Applications in Biomedical Sciences, Shu Ye and Ian N.M. Day

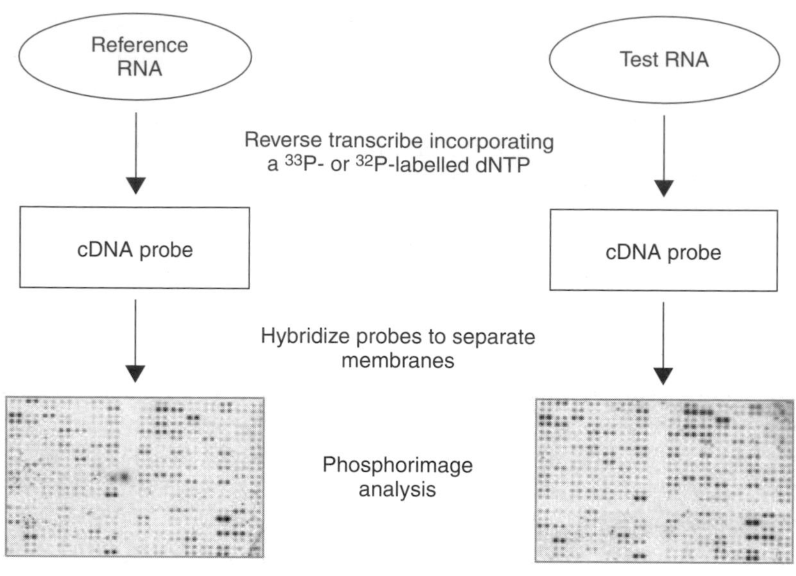

Figure 2.1

Preparation of radioactively labelled probes and hybridization to membrane arrays.

probe, dependent upon the type of array (*Figure 2.1*). For arrays printed on nylon membranes, it is conventional to incorporate a ^{32}P- or ^{33}P-labelled dNTP into the cDNA. ^{33}P is generally preferred because it is a less energetic emitter, reducing the risk of strong hybridization signals interfering with the detection of adjacent signals on the array. The use of the same label to prepare both test and reference probe entails that two identical arrays are hybridized in parallel, followed by detection using phosphorimage analysis. When glass microarrays are used, reference and test probe are synthesized using a different fluor-labelled UTP (*Plate 1*). This has the advantage that both probes can be combined and hybridized simultaneously to the microarray. Following detection of fluorescence using a laser confocal microscope or CCD camera, it is therefore possible to measure the relative level of hybridization between test and reference probe on the same arrayed DNA. Both types of array system have been evaluated and used successfully. However, the microarray format offers three particular advantages over membrane arrays (Schena & Davis, 1999):

1. *Miniaturization*. The solid glass surface used for microarrays is non-porous allowing the deposition of low amounts of material in a defined area. It is thus possible to print arrays at a high density, facilitating a high level of representation within a small area. As a consequence, small reaction volumes are possible, allowing a high concentration of probe and increased hybridization kinetics. This is also aided by the non-porous nature of the support as probe does not have to diffuse into pores to reach arrayed target.
2. *Uniformity*. The solid support provides a uniform attachment surface facilitating a highly defined geometric arrangement of the arrayed targets. This renders the arrays amenable to analysis by digital image

processing procedures, making it feasible to extract expression data in a precise way. In contrast, data extraction from arrays printed on membranes can be influenced by membrane warping, a consequence of the printing, hybridization and washing procedures that can make it more difficult to define the position of arrayed targets. Radionuclide disintegration detection is also influenced by the distance from the source to the detector, such that differences in the warping of parallel membranes can influence the corresponding measurements for the same gene. The ability to compare the amount of reference and test probe that hybridizes to the same arrayed target should also increase the reliability of microarray data.

3. *Speed.* The low reaction volumes mean that hybridization kinetics are relatively fast for microarrays, and the non-porosity of glass allows the removal of unbound probe by washing to proceed more efficiently. Fluorescent detection can also be accomplished within a matter of minutes compared to the days or weeks required to detect bound radioactive probe.

Microarrays can be produced using either oligonucleotide or cDNA representations of the genes. Although the type of expression data obtained is equivalent, there are important distinctions between the two methods with respect to their manufacture.

2.2.1 Oligonucleotide expression microarrays

Oligonucleotide arrays consist of short oligonucleotides (25- to 80-mers) defined based upon known cDNA sequences. The requirement for sequence information means that this type of array can only be used to monitor the expression of known transcripts. However, rapid progress in the analysis of the human genome should entail that this will not be a problem for human transcripts. The choice of which sequences to use requires careful design to avoid regions shared by other genes or complementary sequences in abundant RNAs (Lipshutz *et al.*, 1999). Consideration must also be given to the base composition of the oligonucleotides, to ensure that hybridization conditions are similar across the array. This is particularly important when short sequences are used because the melting temperature (T_m) for such oligonucleotides can be strongly influenced by the base composition. When long oligonucleotides are used (60- to 80-mers) they can be designed with high specificity for individual genes. However, specificity measures may be introduced when shorter oligonucleotides are used by including sequences from several regions within the same transcript or by the inclusion of missense oligonucleotides to evaluate non-specific hybridization (Lipshutz *et al.*, 1999).

Oligonucleotide arrays can be printed by deposition of presynthesized oligonucleotides onto glass slides or more commonly by direct synthesis on the slide by photolithography and solid-phase DNA synthesis. A particular advantage of the latter type of array is that the oligonucleotide is tethered to the surface by one end making it accessible to hybridization, leading to high levels of reproducibility. This can be improved further by incorporating a spacer between the solid surface and the oligonucleotide, rendering all bases accessible to probe (Duggan *et al.*, 1999). Conversely, probe fragments may be long relative to the oligonucleotides, such that a degree of probe fragmentation could be beneficial to improve hybridization (Duggan *et al.*, 1999).

2.2.2 cDNA expression microarrays

cDNA microarrays are produced by spotting PCR-amplified cDNAs onto glass slides. cDNAs represent a much more complex hybridization substrate than oligonucleotides. Although carefully designed amplicons could be generated, in practice cDNAs are amplified from libraries of clones. This carries the risk that arrayed targets may contain regions homologous between different genes, and the possibility that repeat elements are included. However, any clones that show such problems can be identified early in microarray evaluation and replaced or removed if desired. A more significant drawback may be the accessibility of the arrayed cDNA to hybridization. The cDNAs are deposited in a double-stranded form, and probably make multiple constraining contacts with the glass surface (Southern *et al.*, 1999). Heat denaturation is used to separate strands following arraying, but it is not clear how much of the arrayed DNA is rendered in a single-stranded form and available for hybridization. Despite such caveats, the accessibility of cDNA clones of both known and unknown function and the ease of PCR amplification compared with the careful design of oligonucleotide arrays has meant that this type of expression microarray has predominated in the academic field.

2.2.3 Applications of expression microarrays

Expression microarrays can provide two types of information. First, they can be used to catalogue which genes are expressed in a particular cell or tissue sample. For example, this type of analysis can allow the identification of a 'molecular fingerprint' to characterize cells in different stages of differentiation (e.g. Le Naour *et al.*, 2001; Soukas *et al.*, 2001) or at different stages in the cell cycle (e.g. Cho *et al.*, 1998; Spellman *et al.*, 1998). Microarrays can also be used to profile gene expression in diseased tissues, for example different cancers (e.g. Cole *et al.*, 1999; Perou *et al.*, 2000; Sorlie *et al.*, 2001). Second, expression microarrays can be used to study dynamic changes in gene expression over time. In the most straightforward example, expression microarrays can be used to study the gene expression changes that occur when cells in culture are stimulated with a particular agonist (e.g. Heller *et al.*, 1997; Shiffman *et al.*, 2000). However, more complex investigations are also possible to evaluate gene expression changes that may occur *in vivo*, such as in an animal model of disease (e.g. Soukas *et al.*, 2000; Stanton *et al.*, 2000).

A microarray experiment can be divided into several distinct stages, extending from the arraying process, through hybridization, to the downstream analysis of expression data. Below we describe the different stages involved in setting up and using cDNA expression microarrays, including protocols where relevant. Many of the protocols are based upon standard procedures that can be found on the internet at http://www.microarrays.org or http://cmgm.Stanford.edu/pbrown/mguide/index.html.

2.3 Array fabrication

The first stage in a microarray experiment involves the preparation of cDNAs and their arraying onto glass slides.

2.3.1 Obtaining target cDNAs

Several considerations are important when deciding which genes should be represented on a microarray. An ideal would be to include as many different genes as possible as this should increase the power of the analysis in an unbiased fashion. However, logistical considerations include the availability of cDNA clones and the capacity of the glass slide. Three principle options exist for obtaining cDNAs:

1. *Clone sets*. These can be obtained from academic sources or from companies such as IncyteGenomics or ResGen (visit http://www.incyte.com or http://www.resgen.com). Either entire libraries can be obtained or specific clones can be selected. It is important to obtain sequence verified clones and to be certain that clones are provided in a pure state. Both known genes and ESTs can be collected in this way.
2. *cDNA library preparation*. In certain cases, it may be desirable to prepare a comprehensive cDNA library from a particular cell type or tissue. If this approach is taken, individual clones should be isolated and sequenced prior to arraying, to limit redundancy in the arrays. An alternative approach is to prepare a subtracted cDNA library, for example by representational differences analysis (RDA; Nelson & Denny, 1999). This can be useful to limit the genes analysed to those that are actually expressed in a particular tissue or that respond to a particular treatment (e.g. Flores-Morales *et al.*, 2001; Tollet-Egnell *et al.*, 2001). However, it may lead to the exclusion of some interesting genes and may not be so relevant for other experiments. The preselection of cDNAs may also deplete the number of genes that can be used to control the expression data. As with clone sets, both known genes and ESTs can be studied using this method.
3. *Amplification of specific cDNAs*. This is not feasible on a large scale, but may be applicable if only a few hundred genes are to be arrayed. An advantage of this approach is that amplicons can be carefully designed to ensure high specificity.

Whatever approach is taken it is important to include control elements on the arrays. Negative controls include cDNAs from an unrelated species, or cloning vector without insert. Probe hybridization to these clones should be negligible in the final analysis. Positive controls are not so important as many of the represented genes should be expressed in a given experiment. However, cDNAs can be included that are known to be expressed in a particular cell or tissue on the basis of earlier studies. The inclusion of a set of housekeeping genes, the expression of which should not alter between the reference and test condition, can be very important to facilitate normalization of expression data (see *section 2.6.3*).

2.3.2 Preparation of cDNAs for arraying

To prepare for arraying, clones are amplified using vector primers flanking the cDNA inserts. For this purpose it is beneficial to obtain clones prepared using the same vector. In most cases the combination of T7 and T3 or M13 forward and reverse primers can be used. Purified clones or *Escherichia coli* transformants are first deposited into 96-well plates. In the latter case, 96-well plasmid minipreps are required prior to PCR (Mousses *et al.*, 2000). To ensure high yields of cDNA, PCR reactions should be optimized for individual purposes and reaction volumes of 100 μl are used. Standard reaction conditions to amplify cDNAs ranging from 0.2 to 3.0 kb are detailed in *Protocol 2.1*. Following amplification, 2 μl of each reaction is analysed on a 1% agarose gel. This is to check whether amplifications have worked, and to confirm that a unique product is produced. Reactions that have not worked or have low yields may be repeated. If more than one product is produced, this is usually a sign that more than one clone is present in the template. In such cases, it is possible to purify clones following retransformation of *E. coli*. Electrophoresis can also provide an indication of the quantity of PCR products. More informative measurements can be made using fluorimetry (Mousses *et al.*, 2000). To facilitate the printing of high-quality microarrays, final DNA concentrations of between 0.1 and 1.0 μg/μl are recommended (Southern *et al.*, 1999). Although in most cases this should be achievable using a 100 μl PCR, it may be advantageous to set up duplicate 100 μl reactions for each template. Before printing, PCR products are partially purified to reduce unwanted salts, and components of the PCR such as primers. This can be achieved by gel-filtration or precipitation in a 96-well format (*Protocol 2.2*).

2.3.3 Printing PCR products onto glass slides

The printing process involves the sequential transfer of individual PCR products from 96-well plates to defined areas of glass slides. Many types of arrayer are available (Bowtell, 1999). Essentially, they comprise a simple *xyz*-axis robotic arm with a print head that contacts samples in the 96-well plate and deposits the DNA on to the slide. Other features include a place to position 96-well plates and to hold glass slides, a wash station to wash the pens of the print head between different DNA samples, and a computer to run the entire system. Arrayers may differ based upon the number of pens on the print head (4- to 64-pen formats are available), the pen design and the number of slides that can be processed at one time.

The glass slides used for printing are precoated with poly-lysine, amino silanes or amino-reactive silanes (Schena *et al.*, 1996). This serves to increase the hydrophobicity of the slide, improving the adherence of the deposited DNA and minimizing spreading. Protocols are available for performing this treatment (Mousses *et al.*, 2000). However, it may be more convenient to purchase precoated slides that have been quality controlled in their manufacture. Typically, a few nanolitres of each DNA is deposited on to each slide, resulting in the formation of spots in the region of 50–150 μm diameter. The arrangement of spots and the distance separating them on the slide can be programmed into the arrayer, and is determined by the number of clones to be arrayed. A distance of between 200 and 300 μm between each arrayed cDNA is typically used. This will allow the deposition of two arrays on a single glass slide each containing in the order of 3000–6000 cDNAs that can be hybridized under separate 22 × 22 mm coverslips. When fewer than 1000 clones are being printed, sufficient space is available to print replicas of the cDNAs within a single array. This can be very useful to measure variability in the hybridization of probe across an array.

2.3.4 Post-processing of slides

Following printing, DNA is usually cross-linked to the glass slides, and residual amines are blocked by reaction with succinic anhydride (Cheung *et al.*, 1999; *Protocol 2.3*). As a final step, a proportion of the deposited DNA is rendered into a single-stranded form available for hybridization by heat denaturation. This can be done immediately, or slides can be stored in a closed slide box for at least one month prior to post-processing.

2.4 Preparation of labelled probes

Probes are prepared from reference and test RNA pools by reverse transcription reactions that allow the incorporation of a fluorescent tag either directly or indirectly. Following synthesis, probes are separated from unincorporated nucleotides, combined and concentrated prior to hybridization with a microarray.

2.4.1 RNA preparation

The quality and yield of the RNA is of critical importance for the ultimate quality of the probe, and the use of commercial extraction kits is recommended. Protocols that involve multiple precipitation steps are not advisable as these may result in the precipitation of carbohydrates that can ultimately lead to high background on the microarrays (Mousses *et al.*, 2000). Both total RNA and mRNA can be used for probe preparation. The advantage of using mRNA is that probes of higher specific activity may be generated, as random primers can be used in the labelling step. Typical amounts of total RNA required for labelling are 5–20 μg. As little as 100 ng of mRNA can be used. In cases where the amount of RNA is very low, efficient protocols to prepare amplified RNA using T7 RNA polymerase have been described and applied successfully (Phillips & Eberwine, 1996; Luo *et al.*, 1999; Salunga *et al.*, 1999).

2.4.2 Fluorescent probe synthesis

Reference and test cDNA probes are generated that contain a fluorescently tagged nucleotide. The combination of Cy3 and Cy5 UTP has been used most frequently. This is because they are relatively stable in light, can be incorporated efficiently into cDNAs and have a wide separation in their excitation and emission spectra, making it feasible to distinguish both fluors hybridized to the same arrayed DNA. Probe labelling can be achieved either by direct incorporation of Cy3 or Cy5 UTP into the reverse transcription reaction (Schena *et al.*, 1995; *Protocol 2.4*) or by incorporation of UTP containing an allyl amino group followed by subsequent coupling of Cy dye (see http://www. microarrays.org/protocols.html or http://www.operon.com/arrays/array protocols.php). This latter method may be advantageous as the aminoallyl UTP may be incorporated more efficiently than the relatively bulky Cy-labelled UTPs.

2.5 Microarray hybridization

A key advantage of the microarray format is the low reaction volumes required for hybridization, making it possible to use concentrated probe. Hybridizations are performed by dispensing probe under a coverslip and incubating overnight in a sealed hybridization chamber in a water bath. The exact volume of hybridization mix used depends upon the number of arrayed targets and the area covered. Microarrays with fewer than 6000 targets can be hybridized under a 22 × 22 mm coverslip using between 10 and 20 μl of hybridization mix. Larger arrays may require the use of larger coverslips (24 × 50 mm) and 30 μl hybridization mix. As with conventional blotting techniques, pre-hybridization of microarrays is a valuable step for reducing non-specific hybridization (*Protocol 2.5*). For subsequent hybridization, reference and test probe are combined in hybridization buffer containing poly(A) RNA [to block the poly(T) tracts formed when oligo(dT) is used in the reverse transcription reaction] and yeast tRNA (*Protocol 2.6*).

2.6 Generation and analysis of expression data

Gene expression levels are evaluated by measuring the amount of reference and test probe that binds to each arrayed cDNA. Fluorescence is detected on arrays by means of a scanner or reader and saved as a digital image. These data are then imported into software that converts the fluorescence into pixels that can be counted. Following background subtraction and normalization, expression ratios can be calculated.

2.6.1 Scanning

The scanner used to detect fluorescent probe bound to the arrayed cDNAs is essentially a laser scanning confocal microscope (Bowtell, 1999; Schermer, 1999). Separate laser beams of different wavelength are directed onto the slide stimulating distinct emissions from the Cy3- and Cy5-immobilized probes. The resulting fluorescence is detected by an objective lens and stored as a monochrome image. The sensitivity of detection can be adjusted by altering the photomultiplier voltage; in some cases it can be useful to scan arrays at two different voltages as very strong signals may be outside the linear range of detection. Conversely, weak signals may be difficult to detect at low sensitivity. Ensuring that measurements are made within the same dynamic range is important for normalization of data (see *section 2.6.3*). A detailed review of the theory and instrumentation used in the scanning process can be found elsewhere (Schermer, 1999).

2.6.2 Image data generation

The images obtained from the scanner are imported into software that converts the two scans into pseudo-coloured images that can be merged (*Plate 2*). Software for this purpose is available from a number of sources, both commercial and academic (for more information visit http://rana.lbl.gov/Eisen Software.htm and http://www.microarrays.org/software.html). The principles, statistical considerations and protocols for image analysis can be found at these web sites and have been dealt with in detail in several recent articles (Mousses *et al.*, 2000; Baldi & Long, 2001; Bushel *et al.*, 2001; Troyanskaya *et al.*, 2001; Wall *et al.*, 2001; Xia & Xie, 2001). The following describes in simple outline the key events that take place in the extraction of data from scanned images with particular reference to the Scanalyse program (M. Eisen, Copyright ©1998-9, Stanford University).

2.6.2.1 Gridding

The first step in processing an array is to define the location of the different arrayed cDNA spots. This is achieved by overlaying the image with a grid in which each spot is contained within a defined circle. The process of gridding is aided by the highly defined arrangement of spots produced by the robotic printing of the arrays. However, some spots may appear distorted following printing, post-processing, hybridization and washing. It is therefore important that the gridding process be flexible. Once prepared for a single array, grids can be saved and applied to additional arrays from the same batch.

2.6.2.2 Background sampling

Gridding allows the array image to be separated into pixels that are contained within the spot and those that are not. For each arrayed element, a region of defined size surrounding the spot is used for background sampling. This is necessary as the background is not uniform over the entire array. In general, changes in background fluorescence are gradual but may be abrupt in particular regions. In such cases, the intensity measurement of the adjacent arrayed elements may not be reliable. To calculate background, median intensities of the background pixels can be determined separately for the corresponding Cy3 and Cy5 image. Mean values can also be used, but these may be more susceptible to noise (see Scanalyse User Manual at http://www.microarrays.org/software.html).

2.6.2.3 Target intensity extraction

To determine spot intensity, the mean intensity of the pixels contained within the spot circle are determined for the corresponding Cy3 and Cy5 image. Subtraction of the background pixels for the corresponding element then yields a net intensity value for the particular arrayed DNA.

2.6.2.4 Signal filtering

In any experiment, only a subset of the arrayed genes are expected to be expressed. Pixel intensities for unexpressed or poorly expressed genes would be expected to be close to the background intensity for both the Cy3 and Cy5 image. To distinguish such genes, data may be filtered so that only those genes expressed above a certain level are analysed further. Cut-off levels can be defined using various criteria, but typically spot intensities that are less than twofold above background can be excluded. In cases in which the background intensity is abnormally high, spot intensities may be masked and the data from the particular spots may be excluded. In some studies, the percentage of the spot area covered is used as a filtering step, with values being excluded where <40% of the spot area is covered by hybridized probe (e.g. Shiffman *et al.*, 2000).

2.6.2.5 Ratio calculation

The simplest form of analysis is to calculate the ratio of the background-corrected intensity for the Cy5 image to that for the Cy3 image for each spot. In principle, this should equate to the expression level in the test sample relative to the reference sample (where Cy5 has been used to label test and Cy3 to label reference probe). When the corrected values are at least twofold above background, this calculation is reasonable (Schena *et al.*, 1996). However, the accuracy of this measurement can vary, in particular on arrays with high background so most analysis programs incorporate additional estimates of the ratio that can be used for purposes of quality control (e.g. see Scanalyse User Manual at http://www.microarrays.org/software.html).

2.6.3 Processing expression data

To obtain meaningful expression data, the ratios obtained must be processed. Initially, ratios are normalized to exclude any bias towards one of the two probes. Normalized data can then be ranked to identify expression changes between the reference and test condition. Finally, data can be organized to identify similar expression patterns.

2.6.3.1 Normalization

Often the efficiency of label incorporation is not the same for the reference and test probe, ensuring that one of the two probes predominates in the hybridization reaction. Any probe bias must be corrected before meaningful information can be obtained, and to facilitate comparisons between different array experiments. Two types of normalization protocol may be employed.

1. *Housekeeping genes*. All arrays include a set of housekeeping genes, the expression of which is not expected to vary significantly between the reference and test conditions. The ratios for such genes should therefore be close to 1.0. Methods have been developed that extract statistics from the variance of a subset of housekeeping genes that can be used to evaluate the other arrayed genes (Chen *et al.*, 1997). A simple method is to take the mean of the ratios for the subset of housekeeping genes and adjust this to 1.0. The resulting correction factor can then be applied to the other ratios, to obtain normalized data. This can be an effective method when the reference and test sample are closely related, for example, when a single cell type grown in culture is used. It may not be so easy to apply this approach when tissue samples are used, as several cell types are analysed in which the expression of the housekeeping genes may vary. However, array experimentation may be used to identify housekeeping genes for samples of this type in an unbiased way.
2. *Global normalization*. A second approach is to measure the mean intensities for all the genes represented on the array and use this to correct individual ratios. This can be effective because the vast majority of the genes are not expected to change significantly between the reference and test condition. Clearly, the validity of this approach increases with the number of genes represented on the array.

2.6.3.2 Ranking

Corrected expression ratios can be ranked to identify differences between the reference and test conditions. At this stage a threshold must be set to distinguish those genes whose expression does not alter significantly between the two states, and those that are deemed to be significantly up- or downregulated in response to the test condition. This can be accomplished by statistical methods that allow confidence limits to be calculated for the ratio distributions. Typically, expression changes that vary by at least twofold are documented as altering in response to the test state. However, the actual threshold value used can depend upon a number of factors including the nature of the test condition investigated and the volume of data that can be effectively handled.

2.6.3.3 Data mining

Within a single microarray experiment, the ranking of expression ratios allows the identification of genes whose expression alters. However, a particular goal of microarray analyses is to define patterns of gene expression across multiple experiments. The vast amount of data generated demands the development of analysis programs that can organize this data in an unbiased fashion. A particularly valuable technique is to employ cluster analysis (Eisen *et al.*, 1998; Khan *et al.*, 1998; Shiffman *et al.*, 2000). Several types of cluster analysis have been used but the underlying objective is to separate genes into groups that display similar dynamic expression patterns throughout the investigated conditions. A particular value of this approach is that it need not be hypothesis driven, and genes that are involved in different functions that would not previously be associated may cluster within the same group. Combined with biochemical, cell biological and physiological studies, this can provide information on regulatory circuits.

2.7 The need for databases

The use of microarrays demands careful organization of every step of the experimental procedure. It is important to develop databases that hold information on the organization of clones in stored 96-well plates, the arrangement of samples on the microarrays, and the expression data generated. Good databases ensure the smooth flow of information from the recognition of a change in gene expression to the identification of a particular gene. It can also be useful to link specific databases to external databases containing information on a particular gene, such as its function and what is already known of its expression pattern.

2.8 Validation of gene expression changes

As the use of microarrays increases, the reliability of the data generated can be truly evaluated. Without exception, array experiments should be repeated with fresh probe prepared from the same RNAs and the variance between experiments evaluated. It can also be valuable to perform reciprocal labelling such that both Cy3- and Cy5-labelled probes are evaluated for each combination of reference and test RNA. In most cases, the expression of particular genes is quantified using independent techniques (*Figure 2.2*). When sufficient RNA is available, Northern hybridizations or RNase protection assays can be employed (*Plate 3*). Alternatively, real-time quantitative RT-PCR can be used, in particular when only low amounts of RNA are available. When tissue samples have been used, the expression changes observed may be confined to a particular cell type or to cells located at defined regions

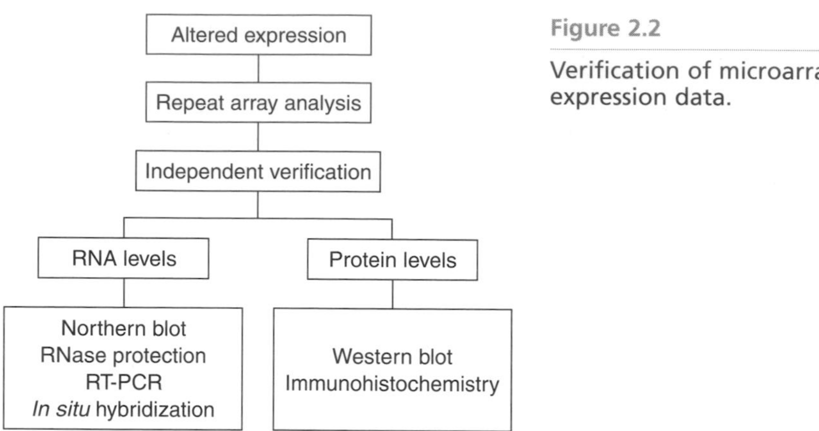

Figure 2.2

Verification of microarray expression data.

in the tissue. In such cases it is valuable to freeze tissue sections at the time of RNA isolation such that mRNA levels or protein levels can be localized by *in situ* hybridization or immunohistochemistry.

2.9 Summary and future prospects

The application of microarray technology to gene expression studies makes it possible to monitor the expression of thousands of genes in parallel. When used in combination with carefully controlled model systems and conventional approaches, there is the real potential to use this technology to begin to dissect complex regulatory circuits.

A future goal would be to develop microarrays that allow all the expressed genes of a given organism to be monitored simultaneously. By extension, this should allow the development of tailored microarrays that could limit the number of genes monitored to those expressed in a particular cell or tissue, or that are involved in a particular disease process. Such developments may require technological advancements such as an increase in the density of arrayed elements that can be printed, and programs to hold and analyse the vast amounts of data generated. In parallel, improvements in the efficiency of labelling and reduction in the amount of starting RNA could be expected. With such developments, it may not be long before it is possible to use expression patterns to 'fingerprint' a variety of disease processes in affected tissues (e.g. Heller *et al.*, 1997; Perou *et al.*, 2000; Sorlie *et al.*, 2001). An increase in the use of microarrays in drug discovery is also anticipated, both in target identification and the assessment of drug efficacy and the evaluation of undesirable effects (Debouck & Goodfellow, 1999).

References

Adams MD *et al.* (1991) Complementary DNA sequencing: expressed sequence tags and human genome project. *Science* **252**: 1651–1656.

Baldi P, Long AD (2001) A Bayesian framework for the analysis of microarray expression data: regularized *t*-test and statistical inferences of gene changes. *Bioinformatics* **17**: 509–519.

Bowtell DD (1999) Options available – from start to finish – for obtaining expression data by microarray. *Nat Genet* **21** (1 Suppl), 25–32.

Bushel PR *et al.* (2001) MAPS: a microarray project system for gene expression experiment information and data validation. *Bioinformatics* **17**: 564–565.

Chabas D *et al.* (2001) The influence of the proinflammatory cytokine, osteopontin, on autoimmune demyelinating disease. *Science* **294**: 1731–1735.

Chen Y *et al.* (1997) Ratio-based decisions and the quantitative analysis of cDNA microarray images. *J Biomed Optics* **2**: 364–374.

Cheung VG *et al.* (1999) Making and reading microarrays. *Nat Genet* **21** (1 Suppl): 15–19.

Cho RJ *et al.* (1998) A genome-wide transcriptional analysis of the mitotic cell cycle. *Mol Cell* **2**: 65–73.

Cole KA *et al.* (1999) The genetics of cancer – a 3D model. *Nat Genet* **21** (1 Suppl): 38–41.

Debouck C, Goodfellow PN (1999) DNA microarrays in drug discovery and development. *Nat Genet* **21** (1 Suppl): 48–50.

Duggan DJ *et al.* (1999) Expression profiling using cDNA microarrays. *Nat Genet* **21** (1 Suppl): 10–14.

Eisen MB, Brown PO (1999) DNA arrays for analysis of gene expression. *Methods Enzymol* **303**: 179–205.

Eisen MB *et al.* (1998) Cluster analysis and display of genome-wide expression patterns. *Proc Natl Acad Sci USA* **95**: 14863–14868.

Flores-Morales A *et al.* (2001) Microarray analysis of the *in vivo* effects of hypophysectomy and growth hormone treatment on gene expression in the rat. *Endocrinology* **142**: 3163–3176.

Hegde P *et al.* (2001) Identification of tumor markers in models of human colorectal cancer using a 19 200-element complementary DNA microarray. *Cancer Res* **61**: 7792–7797.

Heller RA *et al.* (1997) Discovery and analysis of inflammatory disease-related genes using cDNA microarrays. *Proc Natl Acad Sci USA* **94**: 2150–2155.

Hoffmann KF *et al.* (2001) Disease fingerprinting with cDNA microarrays reveals distinct gene expression profiles in lethal type 1 and type 2 cytokine-mediated inflammatory reactions. *FASEB J* **15**: 2545–2547.

Kato-Maeda M *et al.* (2001) Microarray analysis of pathogens and their interaction with hosts. *Cell Microbiol* **3**: 713–719.

Khan J *et al.* (1998) Gene expression profiling of alveolar rhabdomyosarcoma with cDNA microarrays. *Cancer Res* **58**: 5009–5013.

Le Naour F *et al.* (2001) Profiling changes in gene expression during differentiation and maturation of monocyte-derived dendritic cells using both oligonucleotide microarrays and proteomics. *J Biol Chem* **276**: 17920–17931.

Liang P, Pardee AB (1992) Differential display of eukaryotic messenger RNA by means of the polymerase chain reaction. *Science* **257**: 967–971.

Lipshutz RJ *et al.* (1999) High density synthetic oligonucleotide arrays. *Nat Genet* **21** (1 Suppl): 20–24.

Lockhart DJ *et al.* (1996) Expression monitoring by hybridization to high-density oligonucleotide arrays. *Nat Biotechnol* **14**: 1675–1680.

Luo L *et al.* (1999) Gene expression profiles of laser-captured adjacent neuronal subtypes. *Nat Med* **5**: 117–122.

Mousses S *et al.* (2000) Gene expression analysis by cDNA microarrays. In SP Hunt and FJ Livesey, editors. *Functional Genomics: A Practical Approach*, pp. 113–137. Oxford: Oxford University Press.

Nelson SF, Denny CT (1999) Representational differences analysis and microarray hybridization for efficient cloning and screening of differentially expressed genes. In M Schena, editor. *DNA Microarrays: A Practical Approach*, pp. 43–58. Oxford: Oxford University Press.

Okubo K *et al.* (1992) Large scale cDNA sequencing for analysis of quantitative and qualitative aspects of gene expression. *Nat Genet* **2**: 173–179.

Perou CM *et al.* (2000) Molecular portraits of human breast tumours. *Nature* **406**: 747–752.

Phillips J, Eberwine JH (1996) Antisense RNA amplification: a linear amplification method for analyzing the mRNA population from single living cells. *Methods* **10**: 283–288.

Salunga RC *et al.* (1999) Gene expression analysis via cDNA microarrays of laser capture microdissected cells from fixed tissue. In M Schena, editor. *DNA Microarrays: A Practical Approach*, pp. 121–136. Oxford: Oxford University Press.

Schena M, Davis RW (1999) Genes, genomes, and chips. In M Schena, editor. *DNA Microarrays: A Practical Approach*, pp. 1–15. Oxford: Oxford University Press.

Schena M *et al.* (1995) Quantitative monitoring of gene expression patterns with a complementary DNA microarray. *Science* **270**: 467–470.

Schena M *et al.* (1996) Parallel human genome analysis: microarray-based expression monitoring of 1000 genes. *Proc Natl Acad Sci USA* **93**: 10614–10619.

Schermer MJ (1999) Confocal scanning microscopy in microarray detection. In M Schena, editor. *DNA Microarrays: A Practical Approach*, pp. 17–42. Oxford: Oxford University Press.

Shiffman D *et al.* (2000) Large scale gene expression analysis of cholesterol-loaded macrophages. *J Biol Chem* **275**: 37324–37332.

Sorlie T *et al.* (2001) Gene expression patterns of breast carcinomas distinguish tumor subclasses with clinical implications. *Proc Natl Acad Sci USA* **98**: 10869–10874.

Soukas A *et al.* (2000) Leptin-specific patterns of gene expression in white adipose tissue. *Genes Dev* **14**: 963–980.

Soukas A *et al.* (2001) Distinct transcriptional profiles of adipogenesis *in vivo* and *in vitro*. *J Biol Chem* **276**: 34167–34174.

Southern E *et al.* (1999) Molecular interactions on microarrays. *Nat Genet* **21** (1 Suppl): 5–9.

Spellman PT *et al.* (1998) Comprehensive identification of cell cycle-regulated genes of the yeast *Saccharomyces cerevisiae* by microarray hybridization. *Mol Biol Cell* **9**: 3273–3297.

Stanton LW *et al.* (2000) Altered patterns of gene expression in response to myocardial infarction. *Circ Res* **86**: 939–945.

Thimm O *et al.* (2001) Response of *Arabidopsis* to iron deficiency stress as revealed by microarray analysis. *Plant Physiol* **127**: 1030–1043.

Tollet-Egnell P *et al.* (2001) Gene expression profile of the aging process in rat liver: normalizing effects of growth hormone replacement. *Mol Endocrinol* **15**: 308–318.

Troyanskaya O *et al.* (2001) Missing value estimation methods for DNA microarrays. *Bioinformatics* **17**: 520–525.

Velculescu VE *et al.* (1995) Serial analysis of gene expression. *Science* **270**: 484–487.

Wall ME *et al.* (2001) SVDMAN – singular value decomposition analysis of microarray data. *Bioinformatics* **17**: 566–568.

Xia X, Xie Z (2001) AMADA: analysis of microarray data. *Bioinformatics* **17**: 569–570.

Protocol 2.1: PCR amplification of cDNAs

The following protocol has been used successfully for amplifying cDNAs ranging from 0.2 to 3.0 kb.

MATERIALS

Reagents

10× PCR buffer (containing 500 mM potassium chloride, 100 mM Tris–HCl and 1.0% Triton X-100)

25 mM magnesium chloride

T7 and T3 primers

10 mM dNTP mix

Taq DNA polymerase (5 U/µl)

Equipment

96-well PCR plates and plate sealers

Peltier Thermal Cycler

METHOD

1. Aliquot (1 µl) of each cDNA clone into separate wells of a 96-well PCR plate
2. Set up a 99 µl PCR mix as follows:

   74 µl sterile distilled water
   10 µl 10× PCR buffer
   8 µl 25 mM magnesium chloride
   2 µl 10 µM T7 primer
   2 µl 10 µM T3 primer
   2 µl 10 mM dNTPs
   1 µl *Taq* (5 U/µl)

   For each plate a master mix for 100 reactions can be prepared.
3. Using a repeating pipette, transfer 99 µl of reaction mix to the 96-well plate containing the cDNA templates
4. PCR amplify according to the following parameters:

   1× 94°C, 5 minutes
   34× 94°C, 40 seconds, 52°C, 40 seconds, 72°C, 1 minute
   1× 72°C, 8 minutes

Protocol 2.2: PCR product purification by precipitation

MATERIALS

Reagents

3 M sodium acetate

100 and 70% ethanol

3× SSC containing 0.03% sodium lauroyl sarcosine (Sarkosyl)

Equipment

96-well deep-well plates and appropriate centrifuge

METHOD

1. To each well of a 96-well deep-well plate add 10 μl 3 M sodium acetate and 200 μl 100% ethanol
2. Using a multi-channel pipette, transfer remaining PCR products (98 μl) to each well and mix by pipetting
3. Seal plates and leave overnight at −20°C (or for 1 hour at −80°C)
4. Allow plates to thaw and centrifuge at 5000 *g* for 1 hour at 4°C
5. Pour off supernatants and wash pellets by the addition of 200 μl 70% ethanol for 5 minutes
6. Centrifuge at 5000 *g* for 1 hour at 4°C
7. Pour off 70% ethanol and dry plates by inverting over a clean paper towel. Drying can be completed by placing plates in a vacuum oven for 10 minutes
8. Add 40 μl 3× SSC containing 0.03% sarkosyl. In many protocols, 3× SSC alone is used. However, the addition of sarkosyl can improve the morphology of the arrayed DNA. Allow DNA to resuspend by incubating plates in a refrigerator overnight
9. Transfer resuspended DNAs to a fresh 96-well PCR plate. Analyse 1 μl aliquots by agarose gel electrophoresis, to check recovery. Plates can either be used directly to print arrays or sealed and stored at −20°C

Protocol 2.3: Post-processing of glass slides

MATERIALS

Reagents

1-methyl-2-pyrrolidinone

Succinic anhydride

0.2 M sodium borate (pH 8.0)

Equipment

Glass slide holders and jars

Hot plate

UV cross-linker

METHOD

1. Snap dry slides (DNA side facing up) for 3 seconds on a hot plate set at 100°C
2. Place slides in a slide holder, with at least one space between adjacent slides (10 slides is a convenient number to process at one time). UV cross-link DNA to slides using a UV cross linker set at 60 mJ
3. In a fume hood, dissolve 6 g of succinic anhydride in 350 ml of 1-methyl-2-pyrrolidinone. Immediately before use, add 35 ml of 0.2 M sodium borate (pH 8.0) and mix carefully (1-methyl-2-pyrrolidinone is a teratogen and should be handled with extreme care). Transfer slides to solution and agitate for 2 minutes by hand, ensuring that slides do not leave the solution. Shake for a further 15 minutes
4. Quickly transfer slides to a large beaker of water held at 95°C (this can be achieved by switching off the heat to boiling water 5 minutes before required) and agitate by hand for 2 minutes
5. Transfer slides to 95% ethanol for 1 minute. Centrifuge at 700 *g* for 3 minutes to dry slides. Store processed slides in a clean slide box in a dust-free environment

Protocol 2.4: Fluorescent probe labelling

The following protocol describes a generally used method to incorporate Cy dyes directly into the reverse transcription reaction. The protocol employs a commercial kit for the reverse transcription reaction, but efficient labelling can be obtained equally well using other reagents.

MATERIALS

Reagents

CyScribe first strand labelling kit (AmershamPharmacia Biotech)

Cy3-UTP and Cy5-UTP, 25 nmol (AmershamPharmacia Biotech)

2.5 M sodium hydroxide

2 M HEPES-free acid

Equipment

Autoseq G-50 spin columns (AmershamPharmacia Biotech)

Microcon YM-30 filter columns (Millipore)

METHOD

1. In a clean 1.5 ml microfuge tube, set up the following reaction for test and reference mRNAs:

   9 μl 0.1 μg/μl mRNA
   1 μl random nonamer primers
   1 μl anchored oligo(dT) primer

   Mix gently by pipetting and incubate at 70°C for 5 minutes. Allow to cool at room temperature for 10 minutes then briefly centrifuge

2. Place tubes on ice and add the following:

   4 μl 5× CyScript buffer
   2 μl 0.1 M DTT
   1 μl UTP nucleotide mix
   1 μl 25 nmol Cy3 or Cy5 labelled UTP
   1 μl CyScript reverse transcriptase (100 U/μl)

   Mix reactions gently, briefly centrifuge and incubate at 42°C for 1.5 hours.

   Following the addition of Cy-labelled UTP, reactions should be protected from exposure to light as much as possible. In general, Cy3-UTP is used in the reference reaction and Cy5-UTP in the test reaction

3. Degrade mRNA by adding 2 μl 2.5 M sodium hydroxide. Mix and incubate at 37°C for 15 minutes. Neutralize by adding 10 μl 2 M HEPES, mix and briefly centrifuge

4. Apply samples carefully to a pre-spun AutoSeq G-50 column, making sure that leakage does not occur down the sides of the matrix. This can be done conveniently by applying sample in two separate 16-μl aliquots. Centrifuge at 2000 *g* for 1 minute at room temperature

5. Combine probes and make up to 400 μl using Tris EDTA buffer (10 mM Tris–Cl, 1 mM EDTA, pH 8.0). Transfer to a YM-30 centricon column and centrifuge at 10 000 *g* for 7–10 minutes at room temperature. Add 400 μl Tris EDTA buffer and repeat the centrifugation. At this stage, the column should appear almost dry. If a significant amount of liquid is visible, centrifugation should be repeated for 30-second intervals until most of this liquid has flowed through. Invert the centricon column in a clean collection tube and elute probe by centrifugation at 1000 *g* for 3 minutes at room temperature. Typical elution volumes are between 5 and 20 μl. Probe may be used directly for hybridization, or can be stored at −20°C if not required immediately. Probes stored at −20°C for 1 month have been used successfully.

If desired, the efficiency of the labelling reaction can be assessed by direct UV spectrophotometry or by agarose gel electrophoresis followed by fluorescence imaging (see http://www.apbiotech.com/technical/technical_index.html). Alternatively, some of the probe may be used to hybridize to a 'test' array containing a selection of the spotted genes from a larger array.

Protocol 2.5: Pre-hybridization

The following protocol can be used in conjunction with arrays printed in an area within 15 mm^2.

MATERIALS

Reagents

Pre-hybridization buffer (5× SSC, 5× Denhardt's solution, 0.5% SDS, 0.2 μg/μl yeast tRNA, 50% formamide)

Isopropanol

Equipment

Microarray hybridization chamber (Corning)

Plastic coverslips (22 × 22 mm; Sigma)

METHOD

1. Heat 100 μl pre-hybridization buffer at 95°C for 2 minutes, then allow to cool to room temperature
2. Add 15 μl pre-hybridization buffer to each array and cover with a 22 × 22 mm plastic coverslip. Place in a hybridization chamber containing filter discs wetted in water to maintain a humid atmosphere, seal and incubate in a water bath at 42°C for 1 hour
3. Remove the chamber from the water bath and dry the outside thoroughly with a paper towel. Disassemble the chamber and rinse the slide in Millipore water ensuring that the coverslips detach. Transfer the slide to a 50 ml tube containing isopropanol and invert gently five times. Centrifuge at 700 *g* for 3 minutes to dry slide. Proceed directly to hybridization set up

Protocol 2.6: Microarray hybridization and washing

MATERIALS

Reagents

Poly(A) RNA (Sigma)

Yeast tRNA (Sigma)

Hybridization buffer (AmershamPharmacia Biotech)

Equipment

Hybridization chamber (Corning)

Plastic coverslips (22 × 22 mm; Sigma)

METHOD

1. Prepare 16 μl of probe as follows:

   10 μl purified probe
   1 μl of 10 μg/μl poly(A) RNA
   1 μl of 10 μg/μl yeast tRNA
   4 μl of 4× hybridization buffer

2. Incubate at 95°C for 2 minutes, then maintain at 42°C or room temperature before setting up hybridization

3. Pipette 15 μl of probe mix on to the pre-hybridized array, taking care to avoid bubbles. Carefully position a plastic coverslip over the array, again taking care to avoid bubbles but working quickly to avoid probe drying on to the slide

4. Place the slide in a humidified hybridization chamber, seal and incubate overnight in a covered water bath at 65°C

5. Remove the chamber from the water bath and dry the outside thoroughly with paper towels. Keeping the chamber level, work quickly to disassemble the apparatus, transfer the slide to a slide holder and proceed to the first wash step

6. Remove unbound probe by washing as follows:

   1× SSC, 0.03% SDS for 4 minutes
   0.2× SSC for 3 minutes
   0.1× SSC for 3 minutes

   All washes are performed at room temperature with constant agitation.

   Centrifuge slide at 700 *g* for 3 minutes to dry. Scan immediately or protect from light in a slide box if several slides are being processed simultaneously.

Tissue Microarrays: A High-Throughput Technology for Translational Research

3

Susan Henshall

3.1 Introduction

Advances in genomics and the publication of the first draft of the human genome have led to an exponential increase in the number of newly discovered genes implicated in human disease. However, there remains a significant lag period between the discovery and validation of a gene as a clinically useful marker of therapeutic responsiveness or prognosis. This is best exemplified by the limited number of tumour markers that are currently used routinely in the clinical management of patients. For example, oestrogen receptor (ER) and HER-2 expression, which are used to stratify patients for tamoxifen (Early Breast Cancer Trialists' Collaborative Group, 1998) and herceptin treatment (Slamon *et al.*, 2001) for breast cancer, respectively, are two of only a small number in use. This bottleneck can be attributed largely to the laborious process required to assess gene expression in large numbers of patient specimens using conventional methodology. In 1998, Bubendorf and his colleagues at the University of Basel, in collaboration with researchers at the National Institutes of Health, University of Tampere, Finland and Beecher Instruments, first described tissue microarray (TMA) technology (Kononen *et al.*, 1998). TMAs are paraffin blocks containing multiple cylindrical tissue biopsy cores taken from individual donor paraffin-embedded tissue blocks and placed into a recipient block with defined array coordinates. One of the applications of this technology is to accelerate advances in cancer research, by providing a platform for the evaluation of the clinical utility of hundreds of newly discovered genes using cohorts of well-characterized archival specimens, thus providing a potentially more rapid translation to clinical practice. Although the application of TMA technology is not limited to cancer research, most work published to date has focused on the use of TMAs in the assessment of gene expression in different tumour types.

Researchers have sought new technologies such as DNA microarrays (DeRisi *et al.*, 1996) that allow examination of the expression of thousands of genes simultaneously in order to produce a molecular signature for individual tumours. The ability to compare global patterns of gene expression

Microarrays & Microplates: Applications in Biomedical Sciences, Shu Ye and Ian N.M. Day

between samples of cancer and normal tissue and to assess the relationship of gene expression with disease outcome and response to therapy, provides researchers with a rapid method of identifying novel candidate disease- and tissue-specific markers of therapeutic and prognostic utility. TMA technology provides a platform for the efficient evaluation of novel genes identified by DNA microarrays and their involvement at various stages of cancer progression, as well as the assessment of their prognostic and clinical utility using large cohorts of archival material. Consequently, there are an increasing number of studies that apply a two-phase approach, whereby genes that are identified by DNA microarray are assessed subsequently in TMA (Bubendorf *et al.*, 1999a; Moch *et al.*, 1999; Barland *et al.*, 2000; Sallinen *et al.*, 2000).

The widespread adoption of high-density TMAs in many laboratories replaces the one slide–one section approach in which individual archival clinical specimens were placed on separate microscope slides, with the ability to assess RNA, DNA or protein expression in parallel in hundreds of individual patient specimens in a single experiment.

3.2 Tissue microarray construction

The construction of tissue microarrays is shown in *Figures 3.1 and 3.2*. The initial step in TMA construction is the most laborious in the process, but also the most crucial for the quality of the array. This involves identification of the appropriate specimen, and the review of a haematoxylin and eosin (H&E)-stained slide of each specimen by a histopathologist who then marks each H&E slide for the areas of interest (*Figure 3.1*). The complexity of this stage is dependent on the tissue under review, and is often dictated by the

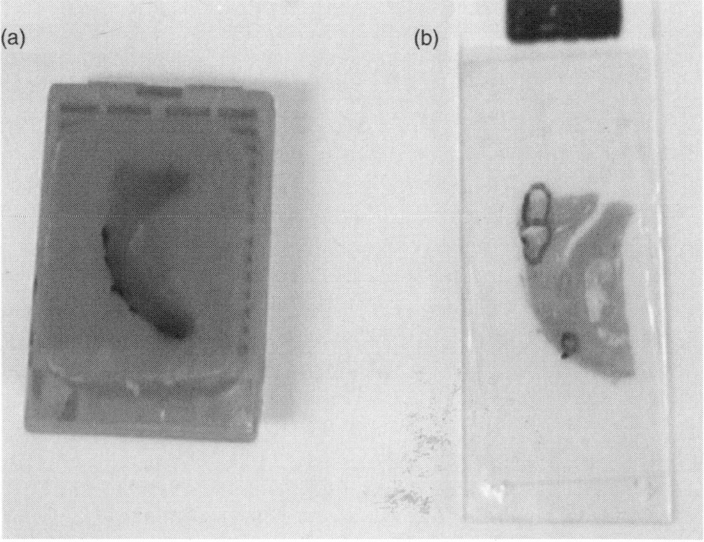

Figure 3.1

The first step in TMA construction from paraffin-embedded clinical specimens. (a) Identification of the appropriate formalin-fixed paraffin-embedded specimen, and (b) H&E-stained slide of each specimen is marked by a histopathologist for the areas of interest.

homogeneity of the tumour type. For example, in the construction of prostate tissue arrays of large, archival cohorts of RP specimens, we routinely review at least 10 slides per case, and each slide is marked for primary, secondary and tertiary Gleason patterns (Gleason & Mellinger, 1974). This approach ensures that we can examine differences in gene expression associated with varying Gleason score as there may be divergent expression of tumour markers associated with varying degrees of tumour differentiation. In addition, we sample the hyperplasia adjacent to cancer and low-grade and high-grade prostatic intraepithelial neoplasia (PIN) if present.

The core tissue biopsies ranging from 0.6 to 2.0 mm in diameter are then taken from individual donor paraffin-embedded tissue blocks of the appropriate specimen and arrayed into a new recipient block (*Figure 3.2*). This is done using a precision instrument developed by the National Institutes of Health and Beecher Instruments (Silver Spring, MD; http://www.beecherinstruments.com) that uses two separate core needles for punching the donor and recipient blocks and a micrometre-precise coordinate system for tissue block assembly. Medium-density arrays consist typically of 1 mm cores taken from ≈100 specimens arrayed in a single block spaced 0.8 mm apart. In comparison, high-density TMAs are typically constructed using 0.6 mm cores of between 400 and 600 individual tissue samples placed ≤0.8 mm apart (Kononen *et al.*, 1998; Hoos & Cordon-Cardo, 2001).

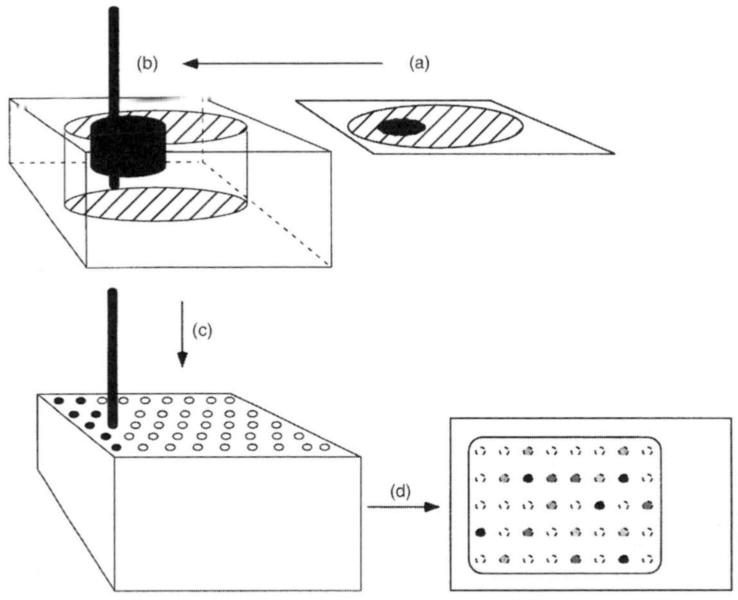

Figure 3.2

Schematic representation of the construction of tissue microarrays.
(a) The histological lesion is identified on the donor paraffin block after H&E staining. (b) Cylindrical tissue biopsy cores (0.6–2.0 mm) are extracted from the donor block and placed into a recipient block with defined array coordinates. (c) The sampling process is repeated to punch different regions in the same donor block and/or hundreds of different donor blocks until a matrix of tissue cores is assembled. (d) The recipient block can be sectioned to produce replica copies of the same tumour set and the array can then be analysed for RNA, DNA and protein by ISH, FISH and IHC respectively, in parallel.

The value of using TMAs is that up to 1000 specimens can be arrayed in each recipient block representing a number of different pathologies and tissue types, both benign and malignant. Depending on the thickness of the original block, between 50 and 400 sections can then be cut from each array and the histology reviewed periodically by H&E staining of a single section.

There are several technical considerations in the optimal construction of high-quality TMAs from paraffin-embedded specimens. Loss of tissue cores during cutting and immunostaining may present a problem. Previously reported rates of tissue loss range from 10 to 30% (Hoos & Cordon-Cardo, 2001), however, a recent evaluation in our laboratory of the rate of loss of tissue cores in medium-density arrays constructed from localized prostate cancers is between 5 and 15% (unpublished data). We have found that tissue loss is reduced when using 1 mm cores compared with 0.6 mm cores of tissue but may ultimately depend on the tissue under study. Tissue loss may also be reduced using adhesive tape systems designed to aid paraffin sectioning (Instrumedics, NJ), as well as placement of tissues containing a high proportion of muscle fibre around the edges of the TMA that preserve the structure of the section. Another technical consideration in the construction of TMAs is defining the orientation of the TMA. This can be achieved simply by placing a core of unrelated normal tissue at the same coordinates of each TMA, for example, placing a core of normal liver tissue at position Row A, Column 1 in each TMA. Hoos and Cordon-Cardo (2001) suggest also placing two orientation cores in specific positions outside the geometric margins of the array to orientate the section after cutting. Previous studies have shown that expression of some proteins may vary between the leading edge of the tumour and the centre of the tumour mass, possibly as a result of the fixation process. Hence, it is advised that cores are taken from both the tumour edge and tumour centre so that adequate representation of the tumour is achieved (Camp *et al.*, 2000). Finally, to ensure uniformity of sections throughout the depth of the TMA it is advisable to utilize paraffin blocks of similar thickness. In many cases, this is addressed by using blocks that have been cut previously only for diagnostic purposes, usually having had only a single section cut for H&E staining. However, where donor tissues are thin, Hoos and Cordon-Cardo recommend punching more than one thin core from the donor block and placing one on top of another in the same location of the tissue array (Hoos & Cordon-Cardo, 2001).

Analysis of gene expression in TMAs depends on the number of cores taken from each tumour block. In arrays in which more than two samples are arrayed per case, it is acceptable to average the staining across the total number of cores represented for each patient. For this reason, as well as to provide adequate representation of each tumour block, many groups support the sampling of at least three cores from each case (Camp *et al.*, 2000; Hoos *et al.*, 2001).

Increasingly, smaller core biopsies are favoured as they enable a number of cores to be taken from the same specimen with minimal damage to the block while preserving histological information. This approach also allows for several replicate arrays to be prepared at the same time. Thus, sections of the tissue array can be used to perform parallel analyses of RNA, DNA and protein expression by *in situ* hybridization (ISH), fluorescence *in situ* hybridization (FISH) and immunohistochemistry (IHC) respectively, on the same tumour set (*Figure 3.3a, b*).

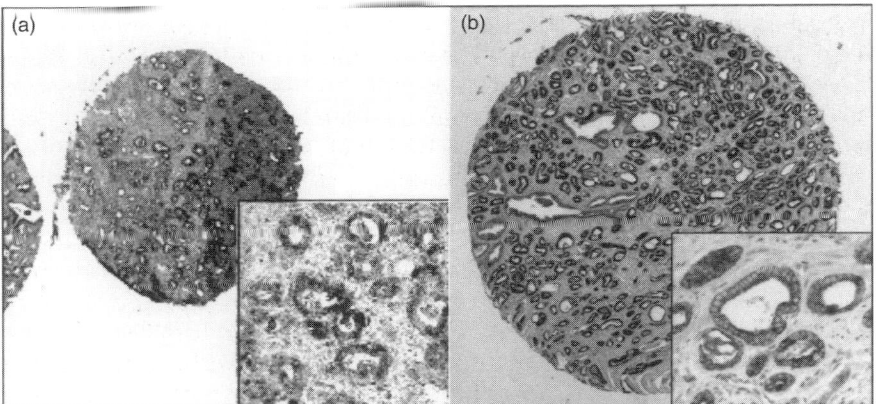

Figure 3.3

Immunostaining of tissue microarrays. (a) Prostate tissue array analysed using *in situ* hybridization for prostate-specific antigen (PSA) RNA. The photograph shows a representative 1 mm core biopsy of a prostate cancer specimen that is strongly positive for PSA. It also illustrates the preservation of the tissue histology even after ISH. The insert is a representative field of the same core at a higher magnification showing PSA-positive epithelium. (b) Representative 1 mm core biopsy stained for a novel molecular tumour marker using IHC (Henshall *et al.*, unpublished). The insert is a representative field of the same core at a higher magnification.

There are numerous advantages of this approach over the conventional one specimen–one slide approach, of which one of the most important is tissue conservation. The issue of tissue conservation is becoming increasingly important with the heavy demand in many institutions for tissue from clinical specimens for use in translational research. The use of TMA ensures the minimum amount of material is removed from each block without compromising the architecture of the remaining tissue, allowing it to be utilized for further diagnostic procedures, such as conventional IHC if required. Also, the limited amount of tissue used in the construction of a single TMA ensures that tissue from the same specimen can be used again in the future.

A detailed cost analysis of staining standard slides versus tissue microarrays demonstrated that the cost of staining 100 standard slides versus one tissue microarray constructed from 100 tumours was not significantly different. This was primarily due to the initial outlay in time and money invested in the construction of the TMA. However, in further experiments, the time and cost were both substantially reduced for use of TMA, with an estimated cost-differential between the use of standard IHC and TMA of approximately threefold (Mucci *et al.*, 2000).

3.3 Staining of tissue microarrays

Sections from tissue microarrays are suitable for the *in situ* analyses of DNA, RNA and protein expression in hundreds of tissues simultaneously. The ability to utilize TMA technology tends not to be limited by the tissues that can by arrayed but rather by the technical problems associated with IHC,

ISH and FISH that are also problems in the analyses of conventional sections. For example, the detection of RNA by ISH in formalin-fixed, archival clinical specimens may be technically challenging, but increasingly new methods of ISH are being established that address these difficulties and can be equally applied to TMAs as to individual slides. Similarly, technical problems may arise in the detection of protein expression by IHC that relate to antigen recognition as a result of formalin fixation, the standard fixative in routine clinical pathology. These can be addressed by experimenting with different methods of antigen unmasking that are common practice in most laboratories experienced in histopathology (MacIntyre, 2001). Improved procedures for the processing and staining of TMAs are now being developed specifically for the use of TMA technology. A recent example is a study that assessed the standard method of FISH versus a novel optimized protocol that was performed entirely on TMAs (Andersen *et al.*, 2001).

FISH, a widely used method to detect chromosomal copy number changes, amplifications, rearrangements and deletions, has been successfully applied to TMAs to detect genetic alterations of several genes in different tissues (Kononen *et al.*, 1998; Bubendorf *et al.*, 1999a, b; Schraml *et al.*, 1999; Andersen *et al.*, 2001). In one of the most comprehensive studies to date, the ability to rapidly screen for gene amplification using a TMA consisting of samples from 17 different tumour types was demonstrated (Schraml *et al.*, 1999). Altogether, 397 individual tumours were arrayed in a single paraffin block and the amplification of three oncogenes (*CCND1*, *CMYC* and *ERBB2*) commonly amplified in human cancers, was assessed. The results demonstrated a 73% concordance between the TMA data and previously reported literature using conventional sections (Schraml *et al.*, 1999). In another large study, the role of cyclin E (*CCNE*) alterations in bladder cancer was investigated using a tissue microarray of 2317 specimens from 1842 bladder cancer patients (Richter *et al.*, 2000). The same TMA was assessed concurrently for cyclin E expression by IHC and the results correlated with patient outcome. The results showed that the majority (62.1%) of *CCNE*-amplified tumours were strongly IHC-positive ($P < 0.0001$), and that low cyclin E expression was associated with poor survival, however, it was not an independent predictor of outcome (Richter *et al.*, 2000). Importantly, this analysis of the prognostic impact of *CCNE* gene amplification and expression in >1800 arrayed bladder cancers was accomplished in only 2 weeks, exemplifying the utility of TMA technology in high-throughput screening of the clinical relevance of tumour markers using standard staining methods. There are now numerous published studies that have utilized TMAs to assess the levels of gene expression in large numbers of tumour samples by IHC (Bubendorf *et al.*, 1999a; Perrone *et al.*, 2000; Richter *et al.*, 2000; Dhanasekaran *et al.*, 2001; Rubin *et al.*, 2001; Torhorst *et al.*, 2001). The major limiting factor for the use of TMA technology as a screening tool for the expression of newly discovered genes by IHC is the design and production of antibodies that are suitable for use in IHC, which currently lags significantly behind the gene discovery process.

The standard method of TMA construction involves the sampling of cores from donor paraffin-embedded tissue, most commonly from formalin-fixed, paraffin-embedded clinical specimens obtained from the archives of pathology departments. However, as already mentioned, some technical problems may exist performing IHC, FISH or ISH using these specimens as a result of antigenic changes in proteins and mRNA degradation due to the fixation and embedding process (Schoenberg Fejzo & Slamon, 2001). In an attempt to address this issue, a recent study assessed the feasibility of constructing frozen TMAs from cores of frozen tissue samples and cell lines embedded in OCT compound (Schoenberg Fejzo & Slamon, 2001). The arrays were constructed using standard 0.6 mm microarray needles (Beecher Instruments, MD), with frozen cores from the donor block being placed directly into an OCT compound recipient block. Sections from the frozen TMAs were then utilized successfully for DNA, RNA and protein analyses, providing a potential alternative to using paraffin-embedded tissue.

3.4 Tissue microarray data and image management

One of the consequences of using tissue microarrays to assess many tumour markers on multiple samples is the voluminous data generated in each experiment. For example, in a single experiment, using a medium-density prostate cancer TMA that incorporates between 2 and 4 specimens of cancer sampled from 150 RP specimens, to examine expression of a single gene by IHC would generate ≈400 separately stained samples. Increasingly, information management systems designed to efficiently archive and manage TMA data and images are being utilized for molecular pathology studies. These information management systems comprise a database, usually implemented on a network server separate from the client computers, and an application that runs on the client computer or through a web interface. The application embodies the analytical requirements of the scientists who use the system. At our institute, a database was implemented using the relational database management system, Sybase. An application, called CanSto was written using the RealBasic development environment. CanSto runs on any PC or Macintosh computer in the network. The database is also available for analysis through a web interface, which is particularly useful for our international collaborators. A general purpose query application (GQA) is used to build and save database queries. GQA can run against all databases designed within the institute's information architecture and gives the researchers the freedom to analyse their data in any way as it constructs standard Structured Query Language (SQL) queries and runs them against the database. The researcher does not need to understand the complexity of SQL as GQA builds and saves the queries for them. GQA allows information on any given gene, histological diagnosis or tissue, separately or in combination, to be retrieved simply.

The CanSto application allows the researcher to collect an H&E image and up to two images of the stained specimen for each experiment performed on the TMA as well as patient ID, array coordinates, array name, block, tissue, tissue source, histological diagnosis and pathology of each specimen arrayed on the TMA. All data entered into CanSto are stored in the Sybase relational database. Images associated with cells of the TMA are stored on a central file server. The files are moderately compressed images in jpeg format. The location of the image file, and the cell that the image is associated with, are stored in the database. The CanSto application works with the Sybase database and the image files to seamlessly present a unified system to the researcher. *Figure 3.4(a–e)* illustrates a sample of the *CanSto* database utilized at our Institute to manage TMA information and histology results from several different projects.

(a) (b)

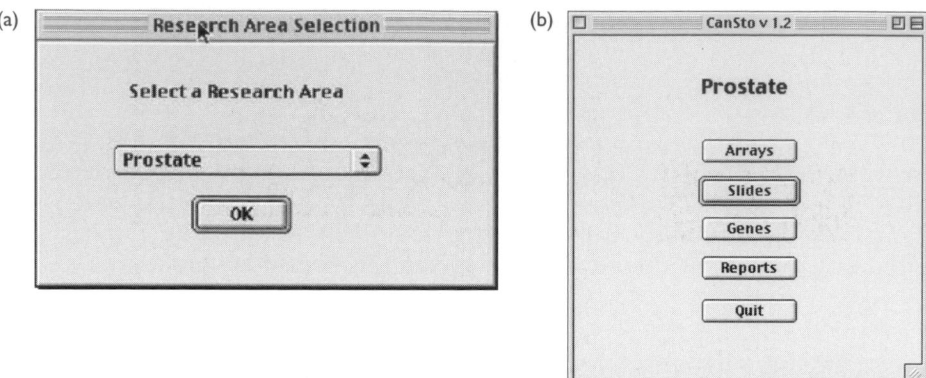

(c)

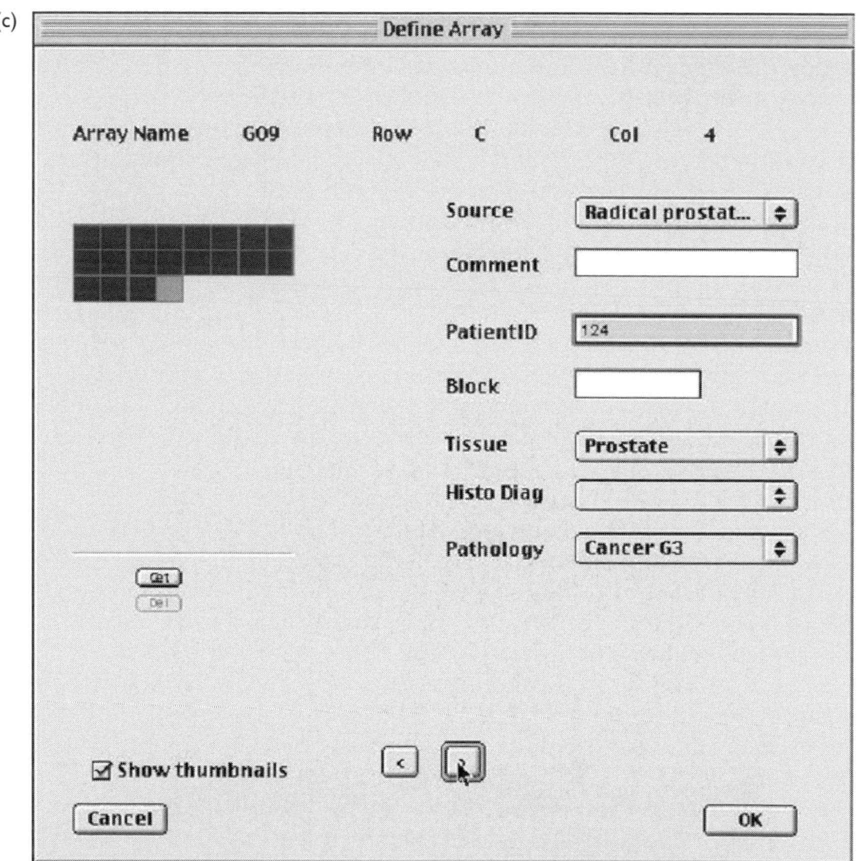

Figure 3.4

The CanSto database to manage TMA data and images for molecular pathology studies. (a) The entry point allows the user to select the research area based on disease site. (b) The user can then choose to enter and edit 'ARRAYS', TMA location information; 'SLIDES', data from individual stained TMA slides; 'GENES', the list of genes for which information is available; or 'REPORTS', reports of histology and patient outcome data accessed via links with disease-specific clinicopathological databases that are managed separately. (c) The composition of an individual TMA is entered only once and includes the pathology details of each sample arrayed in each TMA

(d)

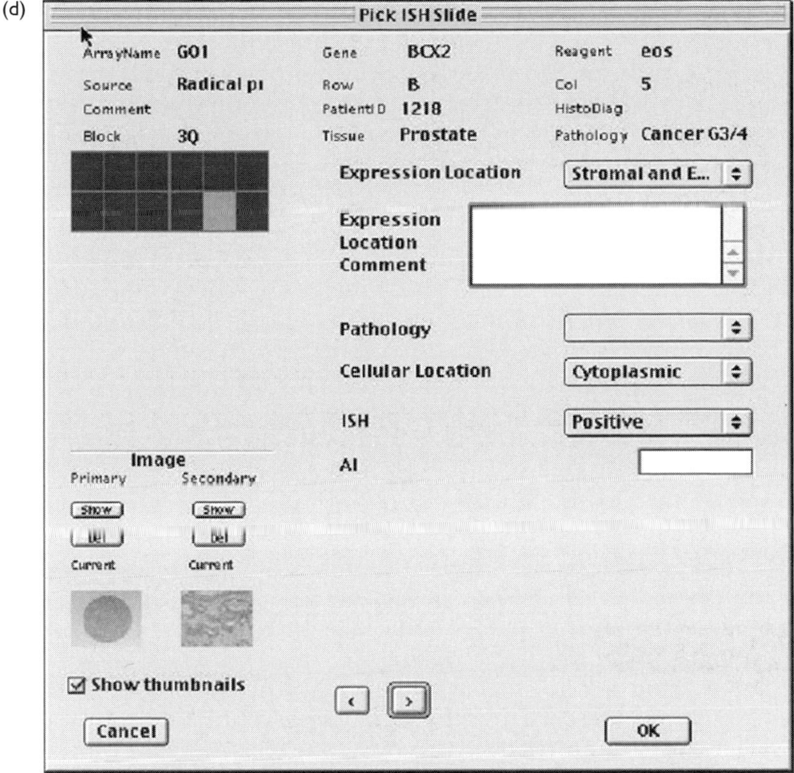

(e)

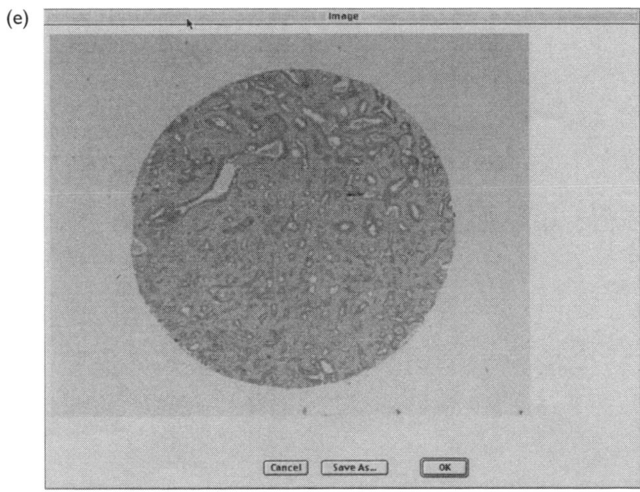

Figure 3.4 *(Continued)*

linked to a representative H&E-stained image of each core. Any data resulting from further experiments performed using this TMA can be entered directly into the database without having to re-enter the details of the TMA itself. (d) The immunostaining results for each stained TMA slide are entered, including the expression pattern and two representative images for each core on the array. (e) Representative image of a stained tissue core. This image can be downloaded from the image database and saved into other applications, for example, for use in presentations or text documents.

It is typical for data in a research institute to be collected using several different mechanisms, often implemented in the prevailing technology at the time collection was commenced. Ideally, unified systems are built to encompass all information management needs, but frequently there is a need to bring together data from separate sources. Our information architecture allows for this. In particular, we have a Filemaker database of disease-specific patient outcome data and other clinical and pathological variables of prognostic relevance from the patients from whom tissue is taken and analysed on the TMA. These data can be brought into the CanSto database and matched automatically by unique patient ID. This allows a GQA query to return not only the TMA data, but also the TMA data matched to clinical outcome details. Thus, the TMA data can be easily correlated with clinical endpoints. The use of unique patient ID ensures that all data accessed in CanSto are anonymous and cannot be traced back to patients.

Another database model was described recently by Manley *et al.* (2001), that utilizes a relational database structure created in Microsoft Access 2000 to manage: (i) clinical and pathology data, (ii) TMA location information, and (iii) web-based histology results (Manley *et al.*, 2001). The TMA component of these databases is comprised of data from 336 prostate cancer patients transferred into 19 TMA blocks with 5451 TMA biopsy cores (Manley *et al.*, 2001). A feature of this model is the integration of a customized imaging system (BLISS; Bacus Labs. Inc., IL and Prostate SPORE Tissue Microarray Working Group) that automatically captures high-resolution composite images of the TMA samples, and assigns a unique name to each image of each sample acquired from the TMA. The images are then linked to an image database that is designed to permit image viewing, entering and editing over the Internet by authorized users, facilitating collaborative research.

3.5 Applications in translational research

3.5.1 Progression model tissue microarrays

Progression TMAs constructed from specimens of different stages in the clinical progression of disease can be used to examine differences in gene expression in different stages of tumorigenesis. For example, published studies have used TMAs to examine changes in gene expression in a progression model of prostate cancer in which large numbers of samples of benign prostate tissue, high-grade PIN, clinically localized prostate cancer and hormone-refractory prostate cancer are analysed under uniform conditions using high-density tissue microarrays (Bubendorf *et al.*, 1999a; Rubin *et al.*, 2001).

The protein, E-cadherin, has been implicated as a potential prognostic indicator in prostate cancer, however, previous studies have yielded conflicting results (Rubin *et al.*, 2001). In a recent study, Rubin *et al.* (2001) exploited the advantage of TMA technology in reducing inconsistencies in staining, to examine E-cadherin expression by IHC in a total of 1220 prostate samples of benign, localized cancer and metastatic prostate cancer lesions arrayed in only 6 high-density TMAs. The study showed conclusively that although aberrant E-cadherin expression was identified in tumours with positive surgical margins, a higher Gleason score and a higher rate of prostate-specific antigen (PSA) failure, there was no statistically significant association with patient outcome. In another study, the differences in gene expression between androgen-dependent and androgen-independent xenograft tissue (human prostate cancer transplanted into nude mice) were analysed by cDNA microarray. Among 5184 genes surveyed, expression of 37 was increased more than twofold in the hormone-refractory xenografts compared with hormone-sensitive xenografts. Of these 37 candidates, the genes encoding insulin-like growth factor-binding protein 2 (IGFBP2) and 27 kDa heat shock protein (HSP27) were consistently over-expressed. To validate these data on prostate cancer specimens, a single TMA of 26 cases of benign prostatic hypertrophy, 208 primary prostate cancers and 30 hormone-refractory local recurrences was constructed and immunohistochemical analysis of IGFBP2 and HSP27 was performed. These analyses revealed a strong association between increased expression of these tumour markers and the progression of prostate cancer to a hormone-refractory state (Bubendorf *et al.*, 1999a). The use of progression TMAs is not restricted to prostate cancer. Bubendorf *et al.* (2001), recently reported the construction of a TMA composed of tissues from 196 node positive breast cancers that included samples from the primary tumour and each of three different metastases as well as samples from 196 node-negative breast cancers (Bubendorf *et al.*, 2001). They report that, using this TMA, they were able to show that high concordance exists between the *HER-2* amplification pattern in the primary tumours and the matched nodal metastases.

3.5.2 Outcome tissue microarrays

Outcome arrays contain samples from tumours with follow-up data. The candidate gene approach, in which the association between a known single gene and the clinical characteristics of the tumour are correlated with disease relapse and outcome, has become commonplace. Several studies have now used TMAs to facilitate the rapid evaluation of the potential clinical relevance of novel genes as molecular targets and markers of prognosis and therapeutic response in different cancers (Bubendorf *et al.*, 1999a, b; Moch *et al.*, 1999; Schraml *et al.*, 1999; Barland *et al.*, 2000; Mucci *et al.*, 2000; Perrone *et al.*, 2000; Richter *et al.*, 2000; Sallinen *et al.*, 2000; Rubin *et al.*, 2001; Torhorst *et al.*, 2001).

Sallinen *et al.* (2000), used the combined approach of DNA microarray and TMA to identify differentially expressed genes of clinical utility in human gliomas. The analysis of pooled glioblastoma samples showed that 107 genes were over-expressed in tumours compared with normal brain tissue. Interestingly, the most outstanding of these was the up-regulation of IGFBP2 as seen in hormone-refractory prostate cancer. Subsequent analysis of IGFBP2 expression in a high-density TMA of 418 brain tumours of varying grade and stage demonstrated a significant association with poor patient survival (Sallinen *et al.*, 2000).

A similar approach was used in a study of new prognostic biomarkers in prostate cancer where analysis of gene expression profiles of normal adjacent prostate, BPH, localized prostate cancer and metastatic, hormone-refractory prostate cancer identified genes that were associated with prostate cancer (Dhanasekaran *et al.*, 2001). Dhanasekaran *et al.* subsequently examined the expression of two of these genes, hepsin and PIM1 by IHC using TMAs constructed from over 300 samples of localized prostate cancer as well as samples of benign prostate tissue, PIN and metastatic disease.

The expression of both proteins was significantly correlated with measures of clinical outcome.

Despite the increasing popularity of TMAs to evaluate molecular targets for their prognostic value, some controversy remains as to their suitability for this use over conventional sections. In an attempt to address this issue, a recent study examined the expression of two well-established prognostic markers, oestrogen receptor (ER) and progesterone receptor (PR) by IHC in a series of 553 breast cancers with comprehensive follow-up (Torhorst *et al.*, 2001). Quadruplicate arrays were constructed by acquiring 0.6 mm biopsies from one central and three peripheral regions of each of the formalin-fixed paraffin-embedded tumours. The results showed that compared with conventional sections, a single sample from each tumour identified ≈95% of the information for ER and 75–81% for PR. However, both the ER and PR analyses performed on conventional sections and TMAs showed a significant association with tumour-specific survival. They concluded that in the case of large series of clinical specimens a single sample from each tumour may often be sufficient to derive information on clinical associations. However, as discussed below, the majority of users of TMA technology promote the use of 2–3 samples from each tumour block (Camp *et al.*, 2000; Hoos *et al.*, 2001).

3.5.3 Multi-tissue microarrays

Multi-tissue microarrays consist of samples from multiple tissues and can include both normal and malignant specimens. These arrays are used commonly to screen several different tumour types for changes in candidate gene expression (Schraml *et al.*, 1999). In addition, the construction of a normal body atlas (NBA) from many different tissues is a valuable resource to comprehensively assess the expression of a newly discovered gene in normal body tissues. This is of particular importance in assessing the suitability of candidate genes as therapeutic targets in order to design drugs with the minimum of non-specific side effects.

3.5.4 Tissue microarrays in animal and experimental models

A largely under-utilized application of TMAs is the assessment of gene expression in animal and other experimental models. For example, tissue from xenograft models or from transgenic or knockout mice can be arrayed and used in the same immunostaining assays as employed for clinical specimens. In our laboratory, we routinely construct tissue microarrays from paraffin-embedded cell pellets of parental and transfected cell lines as well as cell lines treated with different mitogens and growth factors along with the matched control cells. These experimental TMAs facilitate rapid screening of gene expression as well as reducing inconsistencies associated with staining techniques.

3.5.5 Tissue microarrays in clinical trials

A wide application of TMA is in clinical trials. Clinical trial groups such as the International Breast Cancer Study Group have extensive clinical databases with long and detailed follow-up, often with corresponding tumour banks (Coates, 1998; Colleoni *et al.*, 2000). These databases are ideal for evaluation by TMA, as they provide an efficient method for screening gene expression in large numbers of specimens but also allow for the use of very small amounts of tissue.

3.6 Validation of tissue microarrays

Criticisms have been raised against TMA technology that question the validity of this approach. Mostly, these relate to the size of the tissue core arrayed and hence the reduced amount of tissue analysed for each case. It is estimated that a 0.6 mm tissue core, which represents ≈2–3 high-power fields, represents only 0.3% of the area examined in conventional section analysis (Camp *et al.*, 2000). Thus, 0.6 mm tissue samples may not always be representative of the whole tumour and therefore, the prevalence of a molecular alteration may be underestimated (Bubendorf *et al.*, 1999a). However, published data substantiate the use of 0.6 mm cores for homogeneously staining biomarkers (Kononen *et al.*, 1998; Mucci *et al.*, 2000, Richter *et al.*, 2000). Kononen *et al.* (1998), initially described the efficacy of this technique in a study that assessed the frequency of amplification of five genes as well as ER expression in 372 arrayed primary breast cancer specimens by IHC, ISH and FISH. The authors demonstrated that analysis of the ER content from a tissue homogenate by biochemical assay and by IHC from tumour tissue arrays had 84% concordance which is in the range reported in comparisons of biochemical and conventional whole-section IHC ER analysis.

Two recent studies have also shown a high concordance of gene expression detected in TMAs and conventional sections of the same tumour (Camp *et al.*, 2000; Hoos *et al.*, 2001). In the first of these, ER, PR and HER-2 expression were examined in 38 cases of breast cancer, comparing the use of 1–10 cores per case with conventional sections. The results showed that even two cores of each tumour was comparable with analysis of a whole tissue section in more than 95% of cases, and that the degree of concordance increases to 99.5% with five cores per specimen (Camp *et al.*, 2000). In the second study, the protein expression patterns of Ki-67, p53 and the retinoblastoma protein (pRB) were assessed in full tissue sections of 59 fibroblastic tumours and compared with the patterns in one, two and three 0.6 mm biopsies per tumour from the same specimens in an array (Hoos *et al.*, 2001). While the cut-off values for positive and negative samples for Ki-67 and p53 were set at 20 and 10% respectively, the expression of pRB was divided into three categories as has been published previously for the assessment of pRB-positivity in conventional slides (Cote *et al.*, 1998). The results showed that the use of three cores per tumour gave optimal results with concordance rates between tissue arrays with triplicate cores per tumour and full sections of 96, 98 and 91% for Ki-67, p53 and pRB staining, respectively.

Another criticism of TMA technology that relates also to the use of 0.6 mm tissue cores is that it is not suitable for heterogeneous tumour markers that are only expressed in small foci of cells. One example is the tumour suppressor gene *p53*, known to be expressed heterogeneously in a number of different human cancers (Quinn *et al.*, 2000; Sallinen *et al.*, 2000). However, a recent publication that compared conventional whole sections and TMA for *p53* analysis found good agreement between *p53* data from arrayed samples and standard slides (Sallinen *et al.*, 2000).

Despite the limitations of TMA, the potential of this technology to assist high-throughput validation of the clinical relevance of tumour markers is clear and will greatly aid the rapid assessment of expression of new therapeutic and prognostic markers in preclinical trials. It is possible, however, that when assessing the presence or absence of a tumour marker in individual patient specimens in the diagnostic setting, that conventional sections will remain the method of choice.

3.7 Future developments

In the last few years, TMAs have increasingly become a routine tool to screen large numbers of tissue specimens for several molecular markers. Improvements which will facilitate the use of this technology are already underway, with novel automated image scanning and analysis packages being developed by several commercial entities to significantly reduce the amount of manual work that is required. Improvements that are already increasing the overall efficiency of this technique such as the development of database structures that allow sharing of data and images across the Internet are also contributing to its increased use.

Whether TMAs will ultimately replace the use of conventional sections in the diagnostic laboratory is debatable. However, the potential of TMA technology to overcome traditional impediments to the rapid assessment of the clinical relevance of newly discovered genes is clear and it will ultimately facilitate a rapid translation of novel molecular targets for use in clinical practice.

Acknowledgements

We thank Dr Lisa Horvath and Mr Jim McBride for their critical assessment of this manuscript and for the development of the CanSto database. This research was supported by grants from the National Health and Medical Research Council of Australia, New South Wales Cancer Council, the RT Hall Trust, Freedman Foundation, Ronald Geoffrey Arnott Foundation, Australasian College of Surgeons, Australasian Urological Foundation, Prostate Cancer Foundation of Australia and David Wilson Trust.

References

Andersen CL *et al.* (2001) Improved procedure for fluorescence *in situ* hybridization on tissue microarrays. *Cytometry* **45**: 83–86.

Barland M *et al.* (2000) Detecting activation of ribosomal protein S6 kinase by complementary DNA and tissue microarray analysis. *J Natl Cancer Inst* **92**: 1252–1259.

Bubendorf L *et al.* (1999a) Hormone therapy failure in human prostate cancer: analysis by complementary DNA and tissue microarrays. *J Natl Cancer Inst* **91**: 1758–1764.

Bubendorf L *et al.* (1999b) Survey of gene amplifications during prostate cancer progression by high-throughput fluorescence *in situ* hybridisation on tissue microarrays. *Cancer Res* **59**: 803–806.

Bubendorf L *et al.* (2001) Tissue microarray (TMA) technology: miniaturized pathology archives for high-throughput *in situ* studies. *J Pathol* **195**: 72–79.

Camp RL *et al.* (2000) Validation of tissue microarray technology in breast carcinoma. *Lab Invest* **80**: 1943–1949.

Coates A (1998) International Breast Cancer study group trials. *Recent Results Cancer Res* **152**: 429–440.

Colleoni M *et al.* (2000) Early start of adjuvant chemotherapy may improve treatment outcome for premenopausal breast cancer patients with tumors not expressing estrogen receptors. *J Clin Oncol* **18**: 584–590.

Cote RJ *et al.* (1998) Elevated and absent pRB expression is associated with bladder cancer progression and has cooperative effects with p53. *Cancer Res* **58**: 1090–1094.

DeRisi J *et al.* (1996) Use of a cDNA microarray to analyse gene expression patterns in human cancer. *Nat Genet* **14**: 457–460.

Dhanasekaran SM *et al.* (2001) Delineation of prognostic biomarkers in prostate cancer. *Nature* **412**: 822–826.

Early Breast Cancer Trialists' Collaborative Group (1998). Tamoxifen for early breast cancer: an overview of the randomised trials. *Lancet* **16**: 1451–1467.

Gleason D, Mellinger G (1974) Prediction of prognosis for prostatic adenocarcinoma by combined histological grading and clinical staging. *J Urol* **111**: 58–64.

Hoos A, Cordon-Cardo C (2001) Tissue microarray profiling of cancer specimens and cell lines: opportunities and limitations. *Lab Invest* **81**: 1331–1338.

Hoos A *et al.* (2001) Validation of tissue microarrays for immunohistochemical profiling of cancer specimens using the example of human fibroblastic tumors. *Am J Pathol* **158**: 1245–1251.

Kononen J *et al.* (1998) Tissue microarrays for high-throughput molecular profiling of tumor specimens. *Nat Med* **4**: 844–847.

MacIntyre N (2001) Unmasking antigens for immunohistochemistry. *Br J Biomed Sci* **58**: 190–196.

Manley S *et al.* (2001) Relational database structure to manage high-density tissue microarray data and images for pathology studies focusing on clinical outcome. *Am J Pathol* **159**: 837–843.

Moch H *et al.* (1999) High-throughput tissue microarray analysis to evaluate genes uncovered by cDNA microarray screening in renal carcinoma. *Am J Pathol* **154**: 981–986.

Mucci NR *et al.* (2000) Neuroendocrine expression in metastatic prostate cancer: evaluation of high throughput tissue microarrays to detect heterogeneous protein expression. *Hum Pathol* **31**: 406–414.

Perrone EE *et al.* (2000) Tissue microarray assessment of prostate cancer tumor proliferation in African-American and white men. *J Natl Cancer Inst* **92**: 937–941.

Quinn DI *et al.* (2000) Prognostic significance of p53 nuclear accumulation in localised prostate cancer treated with radical prostatectomy. *Cancer Res* **60**: 1585–1594.

Richter J *et al.* (2000) High-throughput tissue microarray analysis of cyclin E gene amplification and overexpression in urinary bladder cancer. *Am J Pathol* **157**: 787–794.

Rubin MA *et al.* (2001) E-Cadherin expression in prostate cancer: a broad survey using high-density tissue microarray technology. *Hum Pathol* **32**: 690–697.

Sallinen S-L *et al.* (2000) Identification of differentially expressed genes in human gliomas by DNA microarray and tissue chip techniques. *Cancer Res* **60**: 6617–6622.

Schoenberg Fejzo M, Slamon DJ (2001) Frozen tumor tissue microarray technology for analysis of tumor RNA, DNA, and proteins. *Am J Pathol* **159**: 1645–1650.

Schraml P *et al.* (1999) Tissue microarrays for gene amplification surveys in many different tumor types. *Clin Cancer Res* **5**: 1966–1975.

Slamon DJ *et al.* (2001) Use of chemotherapy plus a monoclonal antibody against HER2 for metastatic breast cancer that overexpresses HER2. *N Engl J Med* **344**: 783–792.

Torhorst J *et al.* (2001) Tissue microarrays for rapid linking of molecular changes to clinical endpoints. *Am J Pathol* **159**: 2249–2256.

Protein Chips and Microarrays

4

C. David O'Connor and Karen Pickard

4.1 Introduction

The publication of a draft version of the human genome sequence, together with the completion of over 95 other sequences (see http://wit.integratedgenomics.com/GOLD/completegenomes.html), represents a watershed in biological research. From here on, the major challenge will be to characterize the many gene products of currently unknown function that are being uncovered so that it will be possible to develop a truly quantitative understanding of cellular processes.

In principle, one of the most powerful approaches to address this challenge is to identify the ligands that interact with each protein of unknown function. In many cases these will be small molecules of defined structure or protein partners of known function. In other cases, the ligands may be DNA sequences associated with individual known genes. Hence, the principle of 'guilt-by-association' may be used to implicate the unknown gene product in a particular cellular process. Given the scale of the problem, it is becoming clear that the task of identifying ligands needs to be done in a systematic and high-throughput fashion. For this reason, the development of protein microarrays, in which specific polypeptides are arranged as separate addressable elements – and typically spatially separated on a solid substrate – is an attractive concept (Emili & Cagney, 2000).

In addition to allowing the identification of the natural ligands for a particular protein, protein microarrays also have the potential to identify other molecules with the capability to interact with the protein. This possibility is not only of some significance for the development of new therapeutic drugs, but may also impinge on vaccine development. For example, probing a microarray of proteins from a bacterial pathogen with antisera from a patient who has survived the relevant infection may uncover novel antigens with protective properties. The identification of protein ligands also paves the way for more advanced studies to modify the binding specificity. For example, once a substrate of an enzyme has been defined, it becomes possible, in principle, to alter the specificity of the cognate enzyme for biotechnological applications. Although this has traditionally been accomplished by (semi-) rational site-directed mutagenesis, the development of combinatorial approaches, for example, DNA shuffling (Stemmer, 1994; Kolkman & Stemmer, 2001), makes the use of protein microarrays increasingly attractive.

Microarrays & Microplates: Applications in Biomedical Sciences, Shu Ye and Ian N.M. Day

The above considerations indicate that the types and uses of protein arrays are more diverse than their DNA counterparts. Equally, however, the complexity of protein structure and function, and in particular the varied nature of their ligands, means that the development of protein-based arrays is much more challenging than that of gene chips. Despite these difficulties, research in this area is expanding as certain types of information can only be obtained at the protein level. For example, quantification of the level of mRNA transcripts in a sample is not always an accurate indicator of protein abundance (Anderson & Seilhamer, 1997; Gygi *et al.*, 1999). Moreover, the plethora of isoforms specified by open reading frames (ORFs) in a genome may need individual quantification and each may also carry specific post-translational modifications (PTMs), such as phosphorylation or glycosylation. This review summarizes recent progress in preparing different types of protein arrays and also discusses some potential applications.

4.1.1 The importance of miniaturization

Miniaturization of a ligand capture system not only permits the parallel measurement of multiple analytes from small volumes of sample, but also bestows some other significant benefits (Ekins, 1998). For example, from the mass action laws, the degree of ligand bound to a protein ('fractional occupancy') is given by:

$$F^2 - F(1/K[P] + [L]/[P] + 1) + [L]/[P] = 0$$

Where [L] and [P] are the ligand and capture protein concentrations, respectively, F is the fractional occupancy and K is the affinity constant. Importantly, when $[P] \rightarrow 0$, this relationship simplifies:

$$F \approx K[L]/(1 + K[L])$$

Consequently, as [P] decreases, the fractional occupancy becomes independent of the sample volume. In short, the actual concentration of the ligand is not significantly perturbed – and hence can be more accurately measured – if the capture protein concentration is kept low. A further key advantage of miniaturization is that the binding equilibrium is reached more rapidly as the size of the spot decreases, that is, the rate at which the capture protein becomes occupied with a ligand molecule is inversely proportional to the size of the protein spot (Ekins, 1998). Moreover, it is much easier to achieve a uniform distribution of proteins in a spot, and hence a consistent signal, if the spot diameter is reduced.

These attributes of increased sensitivity, speed and uniformity, coupled with reduced amounts of sample and the savings in reagent costs, provide compelling reasons for the development of micro versions of arrays. However, the advantages have to be weighed against the increased demands on the detection methods that are employed to monitor ligand binding and on the major effects of non-specific binding on assay sensitivity.

4.1.2 Types of protein arrays

For the purpose of this chapter, we have classified protein arrays into three groups: proteome arrays, custom protein arrays and protein capture chips (*Figure 4.1*). In the case of proteome arrays, the entire protein complement of a cell type or organism – or as much of it as possible – is gridded onto

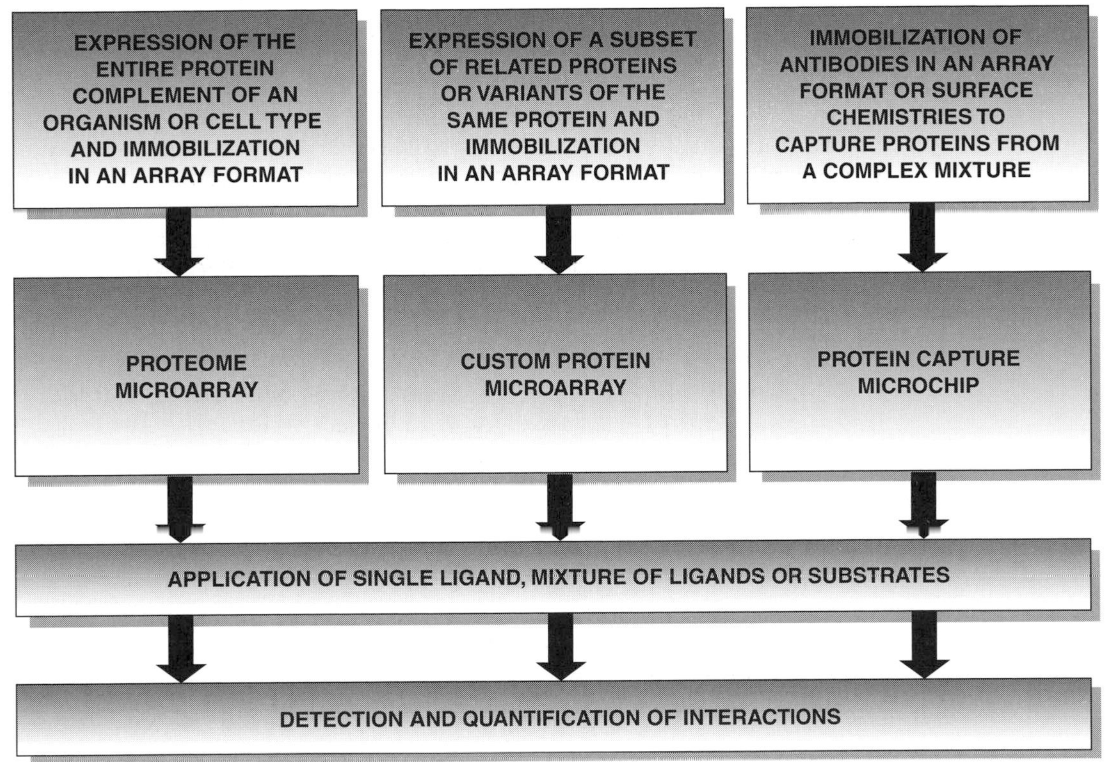

Figure 4.1

Features of three types of protein microarrays.

a solid substrate and screened for interactions with proteins, ligands or substrates and for the existence of PTMs. In contrast, custom protein arrays immobilize a subset of related proteins; for example, proteins that share an enzymatic activity, those involved in a particular biological process such as a signaling pathway or those involved in the pathogenesis of disease. Multiple variants of the same protein may be arrayed, for example, elements harboring different C-terminal truncations or expressing alternative residues at key sites. These types of array can thus be screened to compare the relative affinities of each variant for specific substrates or binding partners. Such systems can also be used to array peptides for epitope mapping and to identify binding motifs or sites of PTMs.

Protein capture chips differ from proteome or custom protein arrays because, unlike the other two categories which may involve the precise gridding of a number of proteins or peptides on a solid support, these arrays may utilize antibodies or specific surface chemistries on the chip to capture proteins of interest from a complex mixture. These types of chips often use mass spectrometry (MS) to characterize the captured proteins. For example, the ProteinChip®, produced by Ciphergen Biosystems, uses derivatized aluminum surfaces to capture specific proteins. These can be subsequently released into the mass spectrometer by surface enhanced laser desorption/ ionization (SELDI-MS) to resolve captured proteins (Merchant & Weinberger, 2000).

Tissue microarrays (TMAs), in which multiple specimens, such as tumour tissue samples, are arrayed in a paraffin block for *in situ* analysis (Nocito *et al.*, 2001), may form a fourth category of protein chip. However, these are not discussed in detail here. Instead we focus on the technology involved in protein microarray production and manipulation as well as the potential applications of the three types of array.

4.2 Microarray technology

Protein microarrays essentially comprise a miniaturized solid support spotted with a protein, peptide or antibody, known as the capture element, in a grid system. The microarray is exposed to another protein or set of proteins or ligands, stringently washed and proteins interacting with the capture agents are detected and identified using specific antibodies or MS. Although there may be variations in this approach, there are five essential steps involved in the preparation and use of a protein array: (i) capture element synthesis, (ii) preparation of the chip surface, (iii) immobilization of the capture element on the chip, (iv) binding of the ligand and (v) detection and quantification of the interaction. These steps are discussed in turn.

4.2.1 Capture element synthesis

Complex mixtures of proteins may be gridded directly onto a solid substrate or proteins and peptides may be produced and purified *in vitro* prior to immobilization. Alternatively, capture elements may be synthesized directly on the solid support.

Heterogeneous protein populations for arraying may be derived from tissue culture medium (Davies *et al.*, 1999), lysed tissue samples (Fung *et al.*, 2001), biological fluids, for example serum or urine (Wright *et al.*, 1999), or bacterial lysates expressing human cDNA clones (Büssow *et al.*, 1998). Proteins may also be affinity purified prior to arraying, by cloning the ORF into an expression vector, inducing expression and isolating the proteins using a tag such as hexa-histidine (His_6) or glutathione-*S*-transferase (GST) (Grayhack & Phizicky, 2001; Zhu *et al.*, 2001). As an example of the latter, Zhu *et al.* (2001), cloned 5800 yeast ORFs into a vector encoding both His_6 and GST tags and expressed these in yeast. Purification of GST-fusion proteins was performed by growing yeast strains in a 96-well plate format, lysing the cells and adding glutathione beads to each well to bind the expressed GST-fusion proteins. In this *tour de force* demonstration of protein expression, they showed that some 80% of the yeast proteome could be expressed and purified.

Peptides may also be synthesized directly onto the solid support (Geysen *et al.*, 1984). For example, the SPOT-technique (Frank, 1992; reviewed by Reineke *et al.*, 2001) involves the deposition of activated amino acids onto a functionalized substrate. The support is treated with an amino acid coupled to a protecting group such as 9-fluorenylmethyloxycarbonyl (FMOC) or nitroveratryloxycarbonyl (NVOC), which may be subsequently removed thereby deprotecting free amino groups (Fodor *et al.*, 1991; Münch *et al.*, 1999; reviewed by Wenschuh *et al.*, 2000). Peptides can therefore be synthesized by repeated spotting of activated amino acids followed by the selective removal of protective groups. Moreover, peptides that have been synthesized in array formats by such techniques can be readily subjected to protein or ligand-binding assays and epitope mapping experiments (Geysen *et al.*, 1984; Münch *et al.*, 1999; Reineke *et al.*, 2001; reviewed by Wenschuh *et al.*, 2000).

Antibody arrays are commercially available, whereby hundreds of different antibodies raised against well-studied proteins are arrayed on a membrane. Antibodies immobilized on protein capture arrays may be monoclonal, antibody fragments or even 'plastic' antibodies. The last of these constitute synthetic polymer particles, which have been engineered to bear antigen recognition sites (Köhler & Milstein, 1975; Haupt & Mosbach, 1998; Borrebaeck, 2000; de Wildt *et al.*, 2000). It is now clear that much of the classical antibody structure is superfluous for binding studies. Thus, antibody fragments or single chain variable fragments (scFvs) can be generated using phage display technology (Clackson *et al.*, 1991). An scFv library can be fused to the N-termini of phage capsid proteins so as to produce a phage display library, whereby each phage presents a unique antibody fragment on the phage coat (Clackson *et al.*, 1991). The phage library itself can then be arrayed (de Wildt *et al.*, 2000).

RNA–peptide fusions have also been synthesized (Roberts & Szostak, 1997) in which an mRNA strand is covalently attached to the protein it encodes. In principle, such fusions could be immobilized on a solid support by anchoring them through a complementary DNA element on an array (reviewed by Zhou *et al.*, 2001).

4.2.2 Chip surfaces and immobilization

The surface of the chip plays an important role in ensuring both the immobilization of the capture element and the reproducible detection of ligand-binding events. The chip surface needs to anchor the capture element securely via covalent or non-covalent bonds so as to reduce the possibility of removal during washing stages and in such a manner so as to prevent protein denaturation. Furthermore, the surface may provide a means of keeping the proteins well separated to reduce cross-contamination. The support may also be tailored to assist the correct orientation of the proteins so that they may readily present binding/catalytic domains to ligands/substrates. Alternatively, the surface may be modified to optimize the detection of bound product (e.g. reduce signal-to-noise ratios). Not surprisingly, therefore, numerous types of derivatized substrates have been explored for arraying proteins (*Table 4.1*).

4.2.2.1 Filters

Early work evolved from immunoblotting techniques, whereby proteins were synthesized or gridded onto filters (Fodor *et al.*, 1991; Münch *et al.*, 1999; reviewed by Wenschuh *et al.*, 2000). The universal protein array (UPA) system described by Ge (2000) comprised 96 spots of purified protein arrayed on a nitrocellulose support. Nitrate groups on the nitrocellulose, which exhibit both hydrophobicity and a negative charge, facilitate protein binding (Gershoni & Palade, 1983; Tovey & Baldo, 1989). The UPA was then probed with a radiolabelled protein of interest and it was possible to quantify the relative affinities of the chosen protein to the arrayed proteins, with the additional advantage of being able to strip and reprobe the filter with further proteins, double-stranded oligonucleotides, or RNA probes (Ge, 2000).

Table 4.1 Characteristics of a number of solid-phase array systems

Array type	Advantages	Disadvantages	Reference
Filter e.g. PVDF, nitrocellulose	Readily derivatized for covalent attachment Cheap Reusable Flexible	Often high background Low capacity as sample tends to spread out	Ge, 2000 De Wildt *et al.*, 2000 Büssow *et al.*, 1998 Lueking *et al.*, 1999 Haab, 2001
Glass	Compatible with standard microarrayers Cheap and readily available Readily derivatized for covalent attachment Low autofluorescence	Possible evaporation Possibility of cross-contamination	Reviewed by Zhu and Snyder, 2001 MacBeath and Schreiber, 2000 Haab *et al.*, 2001 Zhu *et al.*, 2001
Gel pad	Reduced evaporation High sample capacity Reusable	More expensive due to photolithography Possibility of reduced diffusion of larger proteins into gel pad May require longer washing steps	Reviewed by Zhu and Snyder, 2001 Guschin *et al.*, 1997 Arenkov *et al.*, 2000 Vasiliskov *et al.*, 1999
Nanowell, e.g. silicon, PDMS	Cheap Reduced evaporation High sample capacity High signal-to-noise ratio with ^{33}P Easy to apply different wash buffers to each well Silicon chips compatible with MALDI-MS Silicon chips could be integrated into electronic devices	Initial preparation of chip takes time	Reviewed by Zhu and Snyder, 2001 Borrebaeck *et al.*, 2001 Ekström *et al.*, 2001 Zhu *et al.*, 2000

(continued)

Table 4.1 Characteristics of a number of solid-phase array systems (*continued*)

Array type	Advantages	Disadvantages	Reference
Silicon dioxide with silane monolayer	High sample capacity	More expensive due to photolithography Not compatible with standard microarrayers Possibility of cross-contamination	Mooney *et al.*, 1996
Polystyrene	Sheets are compatible with ink-jetting	Often background fluorescence Microtitre plates may require larger amounts of labelled ligand	Silzel *et al.*, 1998 Borrebaeck *et al.*, 2001
Gold/aluminium coated	Gold can be used for SPR studies Can immobilize proteins via cysteine residues	SPR only immobilizes one agent at a time Need to reconstitute capture element after each analyte as applied	Kanno *et al.*, 2000 Williams, 2000 Nelson *et al.*, 2000
ProteinChip®	No protein purification required Can be combined with SELDI-MS and on-chip digestion Wide range of surface chemistries available	Identification can be difficult due to low levels of protein and cross-contamination Proteolytic digestion not always efficient Expensive instrumentation for analyte detection	Merchant and Weinberger, 2000 Fung *et al.*, 2001

The UPA is advantageous over Western blotting techniques because it uses fully active proteins rather than denatured proteins and therefore offers a 10–100-fold increase in sensitivity. Furthermore, the number of interactions analysed in one experiment can be significantly increased.

In contrast to arraying proteins, de Wildt *et al.* (2000) utilized nitrocellulose filters for high-density arraying of antibodies. Antibodies were captured by either coating the filter with antigen or an antibody-binding ligand such as protein A or L. The uses of such antibody microarrays are discussed in *section 4.3.4.1*.

PVDF filters are also highly hydrophobic (Tovey & Baldo, 1989) and can readily bind proteins with a capacity equal to or greater than nitrocellulose. PVDF filters have been used to grid bacterial lysates and even bacterial colonies expressing human cDNA clones for probing with monoclonal antibodies (Büssow *et al.*, 1998, 2000; Lueking *et al.*, 1999). This approach may be modified for spotting lysates of normal or diseased tissue to generate differentially expressed protein profiles (Büssow *et al.*, 1998). However, a disadvantage of using filters as a solid support is that they allow only a limited spot density as each sample tends to spread out (Haab, 2001). Furthermore, filters often exhibit high background levels of signal.

4.2.2.2 Glass

Microscope slides are compatible with standard microarrayer equipment and scanners used for DNA microarrays and have the key advantage of low autofluorescence background. Additionally, and in contrast to nitrocellulose and PVDF, it is possible to achieve smaller spots using glass, with the further advantage of being relatively cheap. Glass surfaces can be coated with a variety of cross-linkers to enable immobilization of the proteins. For instance, a simple approach is to derivatize glass slides with a positively charged poly-L-lysine coating (Eisen & Brown, 1999; Haab, 2001; Haab *et al.*, 2001).

MacBeath and Schreiber (2000), describe two additional types of chemically derivatized glass slides: aldehyde or bovine serum albumin-*N*-hydroxysuccinimide (BSA-NHS)-coated slides. Aldehyde slides, such as those manufactured by TeleChem International, are treated with an aldehyde-containing silane reagent (MacBeath & Schreiber, 2000). Aldehydes react with primary amines present at N-termini and side chains of proteins, resulting in their covalent attachment, in a variety of orientations, to the glass slide. Following printing of proteins the unreacted aldehyde groups on the slide can be quenched using BSA.

BSA-NHS surfaces are fabricated by the modification of gamma amino propyl silane-coated slides (Corning Inc) that exhibit free amine groups on the slide surface. In this case, the amine groups react with *N,N'*-disuccinimidyl carbonate to which a layer of BSA is then attached. The BSA layer is activated with a second layer of *N,N'*-disuccinimidyl carbonate thus amplifying the reactive groups to which surface amines on the arrayed proteins can covalently bind. Such slides can be quenched with relatively small molecules, for example glycine, which facilitates subsequent ligand binding to small capture elements (MacBeath & Schreiber, 2000).

Zhu *et al.* (2001) described the use of nickel-coated slides for the attachment of His_6-fusion proteins to the surface. This type of array may act like a capture array because it is not necessary to purify such a fusion protein prior to immobilization. It is also possible that this type of immobilization may

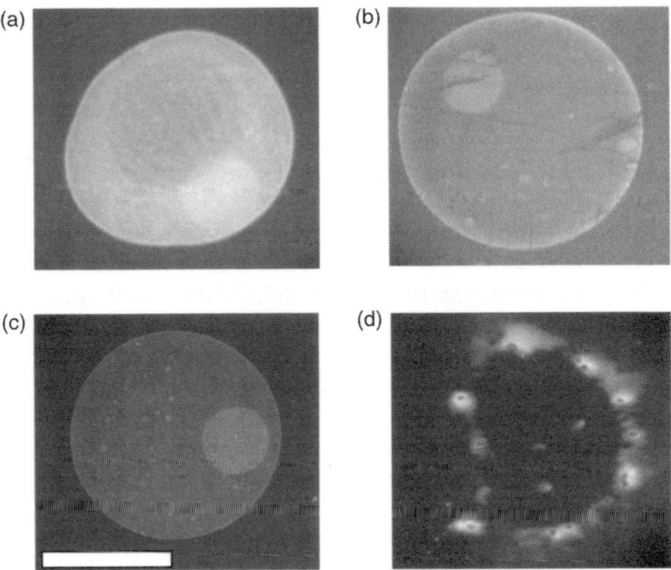

Figure 4.2

Effect of protein concentration and buffer composition on protein distribution in spots and signal intensity. (a–d) Fluorescent microscope images of recombinant GFP spotted manually onto BSA-NHS derivatized glass slides. (a, b) The GFP was prepared in 10 mM Tris–HCl (pH 8.0) at 1 and 0.2 mg/ml, respectively. (c) The same amount of GFP as in (b) but with PBS as the buffer. (d) GFP (0.2 mg/ml) in 20% glycerol/PBS. Note the speckled appearance, indicating aggregation and possible denaturation of GFP. Scale bar: 0.5 mm.

orientate the fusion protein away from the surface of the slide, thereby optimizing its interaction with binding partners.

One of the challenges of using glass slides is evaporation of arrayed proteins. To reduce drying, proteins may be prepared in a glycerol solution (MacBeath & Schreiber, 2000). Such formulations, however, are not without drawbacks. For example, certain proteins, and notably green fluorescent protein (GFP), aggregate in the glycerol/PBS buffer initially used by MacBeath and Schreiber (2000) (*Figure 4.2*). However, they are simple to construct and exhibit good signal-to-noise characteristics for many protein–ligand pairs (*Figure 4.3*).

4.2.2.3 Gel pads

Early work began with the covalent immobilization of oligonucleotides to polyacrylamide gels containing amine or aldehyde groups (Timofeev *et al.*, 1996). Such studies led to the development of a microarray of gel-immobilized compounds on a chip (MAGIChip), which has been used to array proteins and DNA as well as oligonucleotides (Guschin *et al.*, 1997; Vasiliskov *et al.*, 1999; Arenkov *et al.*, 2000). Such samples were immobilized within 3D polyacrylamide gel pads which were typically $100 \times 100 \times 20\,\mu m$ in size. The pads were prepared by photopolymerization of an acrylamide mixture onto a silane-treated glass microscope slide (Guschin *et al.*, 1997; Vasiliskov

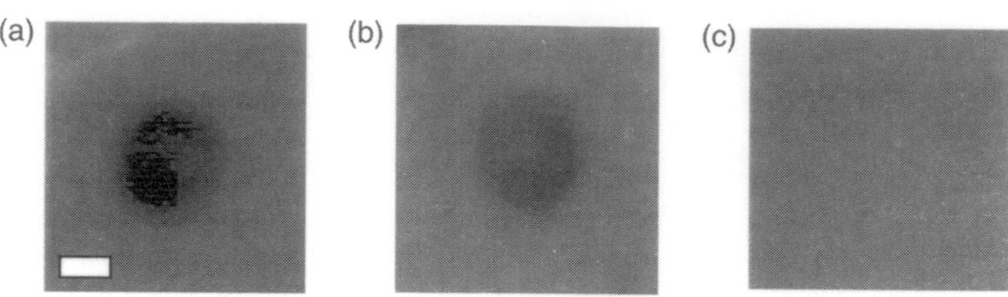

Figure 4.3

Detection of specific proteins in the presence of contaminants using BSA-NHS derivatized glass slides. The figure shows chemiluminescence images captured from BSA-NHS derivatized glass slides spotted with purified $(His)_6$-tagged Ds-Red protein (a), a bacterial cell lysate expressing $(His)_6$-tagged Ds-Red protein (b) or a bacterial cell lysate not expressing $(His)_6$-tagged protein (c). Following the application of an anti-N-terminal-$(His)_6$ antibody conjugated to horseradish peroxidase, spots were detected using chemiluminescence substrates (Pierce). Scale bar: 0.5 mm.

et al., 1999). MAGIChips were then activated by glutaraldehyde treatment. Samples were applied to each well-separated gel pad and the hydrophobic nature of the glass surface prevented cross-contamination. MAGIChips have also been used successfully to perform enzymatic assays and immunoassays and can even be reused following dissociation of antibody–antigen complexes (Arenkov *et al.*, 2000).

Such microchips have a significant advantage over 2D formats such as glass slides in so far as a 3D support provides a higher capacity for immobilization, thereby enhancing the sensitivity of the procedure. A disadvantage of the technique, however, is that the size of the protein or other ligand applied may restrict diffusion into the gel, resulting in stronger signals at the periphery of the gel pad. This challenge may be overcome by the use of different cross-linkers that can improve the porosity of the gel pads (Guschin *et al.*, 1997). Another drawback may be that the gel pad solid supports are more expensive than other types of array systems (Zhu & Snyder, 2001).

4.2.2.4 Nanowells

Nanowells or microwells, essentially miniaturized microtitre plates, are also compatible with standard microarraying equipment and, like the gel pad system, there is the reduced risk of evaporation as the sample is contained within a liquid-filled well. An additional benefit is that the wells keep each protein sample well separated (Zhu & Snyder, 2001). As the chips can be produced in-house, they are also relatively cheap to use (Zhu & Snyder, 2001).

In the best study to date, nanowells of 1.4 mm in diameter were etched in a disposable support by pouring poly(dimethylsiloxane) (PDMS) over an acrylic mould (Xia *et al.*, 1996; Zhu *et al.*, 2000). The PDMS chip (18 × 28 mm) was mounted on a glass slide and a cross-linker 3-glycidoxypropyltrimethoxysilane (GTPS) (Rogers *et al.*, 1999) was attached to the well surface to covalently immobilize the proteins. This system was used to perform a kinase activity assay and it was shown that $[\gamma^{33}P]$ATP binding to the PDMS was reduced relative to microtiter plates, thus improving the signal-to-noise ratio (Zhu *et al.*, 2000).

Borrebaeck *et al.* (2001), produced silicon chips etched with nanowells of 300 × 300 μm in size and a depth of 20 μm. They evaluated their performance with or without a coating of nitrocellulose. Antibody fragments were used as capture agents and after hybridization of applied antigen, the degree of binding was quantified using fluorescence. Interestingly, the nitrocellulose chip produced the highest signal (Borrebaeck *et al.*, 2001). Silicon nanowell chips have also been used as protein capture chips which incorporate MALDI-TOF-MS to observe interacting proteins (Borrebaeck *et al.*, 2001; Ekström *et al.*, 2001). Silicon microchips, presenting captured proteins, were placed onto standard MALDI plates and coated with a thin layer of energy-absorbing matrix solution. The chip surface was scraped to ensure that the matrix was only present in the nanowells. MS was performed and the captured protein was detected with good signal-to-noise ratios.

4.2.2.5 Silicon dioxide surfaces with silane monolayers

Hydrophobic self-assembled monolayers (SAMs) of alkyl silanes such as *n*-octadecyltrimethoxysilane (OTMS) on silicon dioxide surfaces have been used for protein patterning (Mooney *et al.*, 1996). The SAM can be removed from the surface of the chip by UV exposure using a lithographic mask to generate a pattern of exposed and coated SiO_2 on the chip. Proteins absorb preferentially to regions of the chip surface that are coated with OTMS, rather than to those with exposed SiO_2. For example, biotinylated BSA can be adsorbed to the surface generating a 2D pattern of bound BSA. A layer of streptavidin applied to the surface binds non-covalently to the biotin. Subsequently, any biotinylated molecule, such as an antibody can be bound to the streptavidin layer. Mooney and co-workers (1996) have used this technology to immobilize biotinylated goat anti-mouse antibodies in an array format. Fluorescently labelled mouse immunoglobulin was used to probe the patterned surface and binding was detected with a fluorescent microscope.

4.2.2.6 Polystyrene chips

Silzel *et al.* (1998), utilized a polystyrene film of 250 μm thickness onto which ≈80 pl of avidin was dispensed. Covalent immobilization to the film was achieved by derivatization of avidin with a photolabile linker moiety and exposure to a UV light source. Detection of avidin was carried out using fluorescently labelled biotin probes. Antibodies were also immobilized either directly onto the film or onto avidin-printed film via a biotin label. Both avidin and antibodies retained specificity for their targets after drying and were resistant to loss during washing steps. Importantly, the thin sheet of polystyrene was compatible with electrostatic ink-jet dispensing technology whereby a desktop jet printer was modified to deposit spots of solution onto the film.

Polystyrene microtitre plates have also been used as a solid support for arraying proteins. However, 96-well polystyrene plates produced by numerous companies, have been compared and shown to require significantly increased levels of fluorescently labelled antigen to detect interactions with immobilized antibodies, relative to those required by silicon chips (Borrebaeck *et al.*, 2001).

4.2.2.7 Gold chips

Gold-plated surfaces have been used for the attachment of proteins carrying C-terminal cysteine residues. The thiol group of the cysteine residue has a strong affinity for gold and this property can allow the uniform orientation of a protein so as to maximize its interaction with its binding partner. Kanno *et al.* (2000) used this approach to immobilize antibodies. Recombinant protein A, comprising five IgG-binding domains and a C-terminal cysteine, was expressed and assembled on a gold surface. The support was incubated with an anti-IgM antibody that showed high affinity for the protein A on the gold chip. IgM was applied and binding was detected with a peroxidase-labelled anti-IgM antibody. The use of recombinant protein A resulted in increased IgM binding to the anchored antibody relative to that obtained with physically adsorbed antibody molecules. This technology has the advantage of ensuring that the Fab domain of the antibody is displayed to the liquid phase. Chips bearing self-assembled monolayers of alkanethiolates on gold also show promise with homogenous display of peptides resulting in better quantitation and low backgrounds (Houseman *et al.*, 2002).

4.2.2.8 Capture chips

Like silicon chips that have incorporated MS for the detection of interacting proteins, capture chips such as the ProteinChip® (Ciphergen Biosystems Inc.) have combined an array system with SELDI-MS (Merchant & Weinberger, 2000; Fung *et al.*, 2001). ProteinChips® enable the capture, purification and enrichment of specific proteins with similar properties from a biological mixture of proteins. These aluminium chips use a variety of surface chemistries to capture a set of proteins. For example, they can be chemically treated to capture proteins through hydrophobic or hydrophilic interactions, electrostatic interactions (such as cationic or anionic binding) or by metal affinity capture. ProteinChips® can also be customized with biological capture agents such as DNA, antibodies, etc. Up to 16 samples may be applied per chip. Following capture, proteins can be detected by SELDI-MS. The analyte can also be enriched by washing with various stringencies of pH, salt or solvent and subjected to on-chip digestion with proteolytic enzymes (*Figure 4.4*). Subsequent MS analysis of peptides can assist in the identification of captured proteins, using the principle of peptide mass fingerprinting (*section 4.2.5*).

4.2.3 Sample deposition

The technology adopted for applying the protein to the chip depends to some extent on the type of solid support chosen. With the increased use of DNA arraying tools it may be appropriate to utilize the same robotic spotters for arraying of proteins. However, there are now alternative systems designed to generate high-density protein arrays that are more compatible with alternative supports.

4.2.3.1 Microcontact printing

Lueking *et al.* (1999) used an automated spotting robot equipped with a transfer stamp comprised of 16 pins (250 μm tip size) to array proteins onto PVDF filters. It was possible to deposit >100 000 samples on a filter of

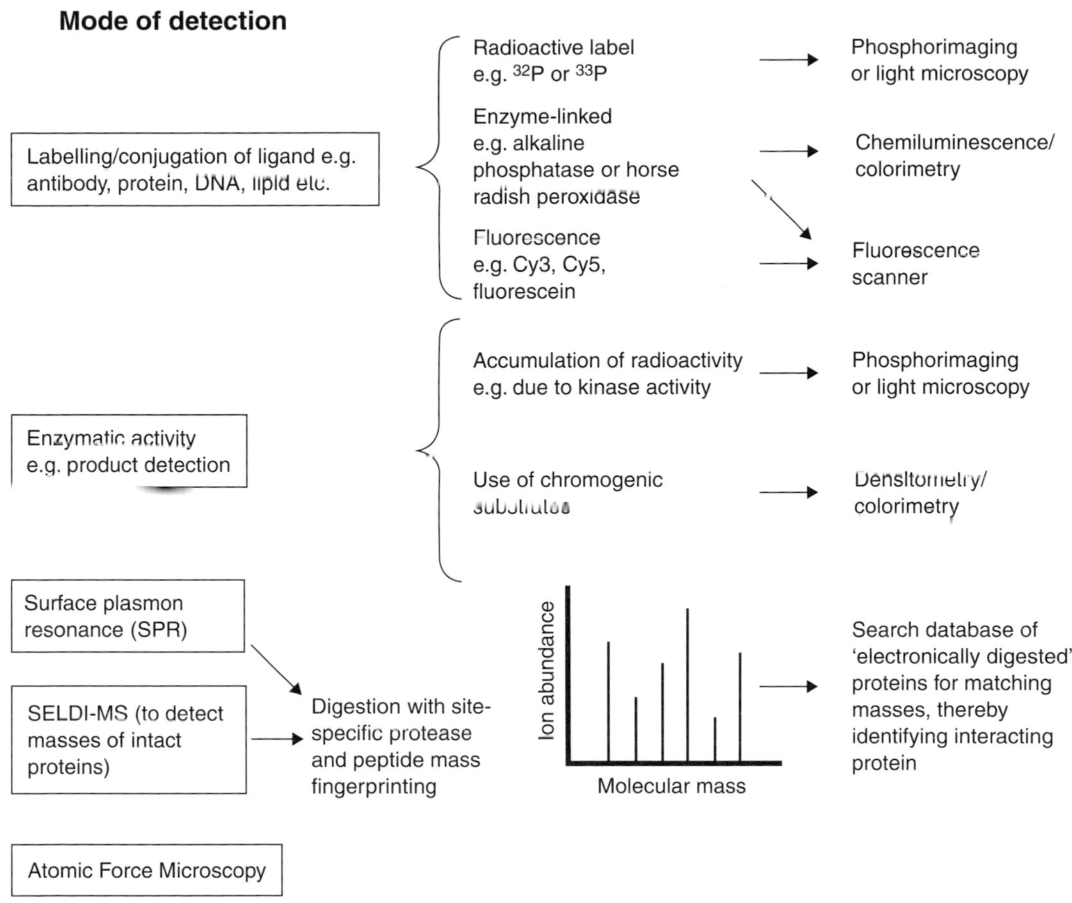

Figure 4.4

Detection of ligands interacting with protein microarrays and microchips. The interaction between the capture element or binding chemistry on the chip and its ligand(s) may be detected by a number of common methods. The chip may be probed with a ligand labelled with a fluorescent dye, radioactivity or conjugated to an enzyme. Alternatively, if the array is to be screened for enzyme activity, the product of the activity may be detected following substrate application. Binding and dissociation events may also be observed using atomic force microscopy or SPR. The latter can be combined with proteolytic digestion and mass spectrometry to identify interacting partners. Certain types of protein capture chips can also utilize SELDI-MS to detect the masses of proteins associated with the chip. Digestion with a site-specific protease and peptide mass fingerprinting can subsequently be used to assist in their identification.

222×222 mm and, when miniaturized to the size of a microscope slide, the filter could accommodate 4800 samples. Similarly, MacBeath and Schreiber (2000) used a high-precision contact printing robot from Affymetrix as well as an in-house built split-pin arrayer comprising 32 pins. In this case, it was possible to print 10 800 proteins in half the area of a standard 2.5×7.5 cm glass microscope slide.

An alternative manual spotting device was described by Guschin *et al.* (1997). This group used a thermostabilized, gold-plated glass fibreoptic pin of 240 μm in diameter to apply ≈1 nl of solution (this group spotted mouse IgG1, rabbit IgG and BSA) onto polyacrylamide gel pads. The chip and the single pin were placed under a binocular microscope and positioned using manual manipulators.

4.2.3.2 Microdispensers

Deposition of droplets of sample has been performed using various micro-dispensing technologies. Electrospray deposition involves depositing charged protein or DNA solutions onto a substrate through holes in a dielectric mask (Morozov & Morozova, 1999). Spot deposition is controlled by a local electrostatic field that attracts the sample to the substrate through the arrayed holes. Substrates include any slightly conductive surfaces such as aluminium or gold plating, nitrocellulose or PVDF membranes. Protein spots of 2–6 μm can be deposited with retention of catalytic activity or specificity of antibody binding. Electrospray deposition also has the advantage over other techniques of a fast delivery rate with a high efficiency of transfer (Morozov & Morozova, 1999).

Borrebaeck *et al.* (2001) developed an in-house piezoelectric flow-through dispenser for placing 100 pl droplets into nanovials. The chip, mounted on a stage, could be manoeuvred with high precision following the deposition of the desired number of droplets into each vial. Multiple deposition of a dilute protein solution in a small defined surface area has been shown to enrich the sample and is described as 'spot-on-chip' technology. The desired enrichment factor is determined by the number of droplets deposited. The increased surface density of analyte in the nanovial is desirable for sensitive MALDI-TOF analysis (Ekström *et al.*, 2001) and may be 10–50 times as sensitive as a single droplet of sample. High-speed picolitre dispensing has also been used to spot liquid expression cultures onto PVDF filters (Walter *et al.*, 2000).

Ink-jetting was described by Silzel *et al.* (1998), who disassembled a print head unit of a desktop ink-jet printer and loaded it with the protein sample. Using the adapted printer they were able to print arrays of 100 pl of avidin or antibody spots of ≈100 μm in diameter, onto sheets of polystyrene film (Silzel *et al.*, 1998). Commercial ink-jet dispenser systems (e.g. the BioChip Arrayer, Packard Instrument) are also available and have been employed to deposit protein samples onto MAGIChips (Vasiliskov *et al.*, 1999).

4.2.3.3 Lithography

Photolithography involves the light exposure of a solid support derivatized with photolabile protecting groups, via a mask. This type of approach has been used for the synthesis of peptides onto a solid support. For example, a glass slide with amine groups on its surface can be derivatized with a photolabile protecting group such as NVOC. The slide can then be exposed to light through a mask that effects deprotection in exposed regions leaving reactive amines. The support can subsequently be treated with an amino acid (also bearing a photolabile group) that readily reacts with the deprotected regions of the slide. The support can be illuminated again to remove the photolabile group from the attached amino acid and the cycle repeated to generate different short peptides simultaneously (Fodor *et al.*, 1991). Photolithography could therefore be adapted to immobilize proteins in an array format onto an NVOC-coated slide, for example.

4.2.4 Application of ligand/substrate

Purified proteins, protein mixtures, antibodies, DNA or enzyme substrates are commonly used to interrogate protein microarrays. Such solutions may be incubated with or passed over the chip, as in biosensor systems. The pH of the solution may be critical for the interaction between the captured agent and binding partners and may not necessarily be compatible with every binding pair. Washing reagents such as H_2O, Tris- or phosphate-buffered saline with or without 0.1% Tween-20, have been used to remove non-specifically bound ligands (Guschin *et al.*, 1997; Lueking *et al.*, 1999; MacBeath & Schreiber, 2000; Borrebaeck *et al.*, 2001; Zhu *et al.*, 2001).

In the case of capture chips to which complex protein mixtures are applied, the buffering conditions can be used to bind proteins with certain characteristics selectively. Furthermore, the stringency of the washing buffers can determine what remains bound to the chip. For example, identical spots of sample can be washed with a buffer of increasing ionic strength. It is possible to enrich proteins in this manner to near homogeneity (Merchant & Weinberger, 2000). In all circumstances, success is largely dictated by the affinity constant of the protein–ligand interaction. As a rule of thumb, it is unlikely that interactions of less than nanomolar will be stable to most washing steps.

4.2.5 Detection of interactions

Several techniques have been adopted to allow the sensitive detection of interactions between an arrayed element and an interrogating ligand. Some of these are described below (*Figure 4.4*). In each case, there is a trade-off between the sensitivity of detection and the specificity.

Fluorescent dyes such as Cy3 (green), Cy5 (red), BODIPY-FL (blue), FITC (green), rhodamine (red) and Texas Red, are a sensitive means of detecting interactions between captured elements and ligands (Mooney *et al.*, 1996; Guschin *et al.*, 1997; MacBeath & Schreiber, 2000). Such dyes can be conjugated to antibodies or protein samples and used to probe the microarray. Fluorescence slide scanners fitted with cooled CCD cameras, together with a magnifying lens and appropriate excitation and emission filters (Guschin *et al.*, 1997; MacBeath & Schreiber, 2000; Walter *et al.*, 2000) or alternatively, laser confocal scanning devices, may be employed (Bowtell, 1999). Mixtures of differentially labelled protein samples can be applied to a chip and fluorescence from each detected using different specific filter sets. The ratio of fluorescence between the two channels is indicative of the relative concentration of each protein sample, thereby enabling comparable quantification of protein binding (Haab *et al.*, 2001). However, a disadvantage of using fluorophores to label ligands, is that the size of the fluorescent dye may affect the structure of the protein (Haab, 2001).

Radiolabelling of ligands does not greatly interfere with the conformation of proteins and therefore offers an alternative method of detection. For example, Ge (2000) labelled the fusion protein GST-K–p52 with [γ^{32}P]ATP using heart muscle kinase and probed an array comprising transcription factors, activators and co-activators to detect novel interactions by autoradiography and densitometry.

Radioisotopes have also been used to investigate kinase activity (MacBeath & Schreiber, 2000; Zhu *et al.*, 2000; Zhu & Snyder, 2001). A selection of substrates and [γ^{33}P]ATP were applied to an array of potential yeast protein kinases, followed by washing and resolution by phosphorimaging (Zhu *et al.*, 2000). Conversely, glass slides spotted with kinase substrates were incubated with [γ^{33}P]ATP and various kinases (MacBeath & Schreiber, 2000). Kinase activity was visualized by developing the slides directly and observing the presence of grains of silver on the slide with an automated light microscope (MacBeath & Schreiber, 2000).

Ligands, including antibodies and protein A or L, may be conjugated to enzymes such as horseradish peroxidase or alkaline phosphatase, and ligand binding can be visualized by the application of an enzyme substrate. For instance, chemiluminescence is used to detect binding of horseradish peroxidase-conjugated ligands (Huang, 2001) and alkaline phosphatase-conjugated ligands can be visualized by fluorescence using the substrate 2-(5′-chloro-2′-phosphoryloxyphenyl)-6-chloro-4-(3H)-quinazolinone (ELF®97 phosphate, Molecular Probes) or with 5-bromo-4-chloro-3-indolyl phosphate/Nitroblue tetrazolium (BCIP/NBT) substrate to produce colored products (Mendoza *et al.*, 1999; Arenkov *et al.*, 2000; de Wildt *et al.*, 2000; Avseenko *et al.*, 2001).

Detection of protein interactions has been significantly improved with the development of on-chip MS. As mentioned earlier, the SELDI ProteinChip® captures proteins directly onto a derivatized aluminium chip, and can detect captured proteins by laser desorption/ionization time-of-flight mass analysis (Fung *et al.*, 2001). After washing off non-specifically bound proteins, a UV light energy-absorbing matrix is applied and a laser beam focused on the sample, thereby desorbing and ionizing proteins into the accelerating electric field of a MALDI-TOF mass spectrometer. The ionized proteins 'fly' according to their mass/charge ratios. Hence, by measuring the time-of-flight to the ion detector and deconvoluting to a charge state of 1, the mass of the protein can be deduced (Merchant & Weinberger, 2000). *In situ* digestion with a site-specific protease followed by MS to measure the peptide masses and subsequent database searching (peptide mass fingerprinting) is a powerful approach for the identification of captured proteins, but relies heavily on the presence of the predicted protein sequence being present in a protein sequence database. However, strategies have also been developed for *de novo* sequencing by MS (Merchant & Weinberger, 2000). The major current drawback with the MS-based detection approach is that it is not quantitative. This is mainly due to the variable efficiencies at which proteins are ionized in complex mixtures.

Surface plasmon resonance (SPR) can also be used to investigate binding and dissociation kinetics of an anchored analyte and a solution-borne ligand. For SPR studies, the capture element is bound to 1×1 cm gold-plated chips and inserted into a biosensor, for example, a Biacore AB instrument. The ligand is passed continuously over the chip and interactions are observed in real-time. SPR biosensors measure changes in the refractive index in the aqueous layer close to the surface of the chip caused during complex formation or dissociation. Such changes are relative to the amount of ligand bound to the chip (Williams, 2000). As SPR is non-destructive, it can be coupled with MS to characterize the protein bound to the chip (Merchant & Weinberger, 2000; Nelson *et al.*, 2000). Captured ligands can be eluted and directed to an active flow cell where proteolytic digestion occurs (Nelson *et al.*, 2000). Peptide mass fingerprinting is then used to assist protein identification.

Atomic force microscopy is a further detection technique that can measure the change in height of a surface-bound element following ligand coupling, to quantitatively assess protein interactions (Jones *et al.*, 1998; Silzel *et al.*, 1998). Alternatively this procedure may be used to measure changes in surface roughness, which will reflect the presence of a bound ligand.

4.3 Versatility of protein microchips

4.3.1 Constraints

In an ideal array system, the principle is to array pure proteins on a solid support via specific and stable linkages, retaining protein conformation and function. Each protein would need to be presented in its entirety, including appropriate PTMs, with readily accessible binding domains or active sites. The array could then be probed with a single protein/ligand or mixture of proteins/ligands that would interact specifically with its target and be detected with high sensitivity. Such interactions could be accurately quantified and comparable.

In reality, it may not be possible to isolate each protein in a pure form because affinity columns frequently also pull down non-specific proteins (see e.g. Owens *et al.*, 2001). Moreover, membrane proteins create a particular challenge because they may require the presence of a detergent or phospholipid to aid solubilization (Zhou *et al.*, 2001). Furthermore, the presence of a tag, necessary for affinity purification, may alter the folding properties of the protein thus reducing its affinity for binding partners or substrates. Attachment of a protein to a solid support may also mask the active site or binding domain so that it is not optimally presented to the ligand. It is also possible that the surface chemistry of the chip is not suitable for immobilization of every protein onto the array surface with equal efficiency. Similarly, the pH of the binding and wash solutions may favour interactions between some sets of proteins and not others and may even be too stringent, thus removing weakly bound ligands.

Some *in vitro* systems are also unlikely to reproduce PTMs that may occur *in vivo* and that may be necessary for protein function. Likewise, particular cofactors or components of a complex, that may be essential for enzymatic activity or protein binding, may not be present in such a system, resulting in false negatives. Finally, the labelling of each probe may not be equally efficient and detection methods can often produce false positive or negative results due to cross-reactivities and non-specific interactions. Despite these potential drawbacks, early results have been encouraging and are already uncovering important and unanticipated new protein–ligand interactions.

4.3.2 Pioneering studies

Several workers have investigated the limits of arraying proteins using a variety of solid supports. MacBeath and Schreiber (2000) explored the versatility of glass microscope slides by choosing three proteins; protein G, p50 (of the NK-κB complex) and the rapamycin-12 kDa FK506 binding protein (FKBP) domain of the FKBP-rapamycin associated protein (FRAP), each of which had an interacting partner, and spotting all three, in quadruplicate on aldehyde-coated glass slides. Each slide was exposed to a fluorescently labelled binding partner. Therefore, when the slide probed with BODIPY-FL-conjugated IgG was scanned, only the spotted protein G was detectable. Likewise, a slide probed with Cy3-IκBα only illuminated p50 spots and, when probed with Cy5-FKBP12 in the presence of rapamycin, only FRB spots were detected. In fact, to confirm the specificity of these interactions, when 10 800 spots of protein G and a single spot of FRB were printed on a slide and probed with both BODIPY-FL-conjugated IgG, and Cy5-FKBP12 with rapamycin, it was possible to pin-point the single spot of FRB by its red fluorescence signal amidst the blue protein G signals. This group also demonstrated the exclusive interaction between immobilized proteins DIG, biotin and synthetic pipecolyl α-ketoamide and small ligands anti-DIG antibody; streptavidin and FKBP12 each coupled to BSA labelled with different fluorophores (MacBeath & Schreiber, 2000).

The interaction between antigen and antibody pairs was investigated further by Haab *et al.* (2001). One hundred and fifteen antigens were immobilized on poly-L-lysine-coated slides and probed with fluorescently labelled antibodies of increasing concentrations. These interactions were detected and quantified by measuring the fluorescence signals. They found that antibody labelling was highly efficient, possibly due to the relative stability of antibodies and the uniform nature of the Fc portion, which was labelled at amine groups on lysine residues. As a result, it was possible to detect and accurately quantify the interaction between 50% of the antigen–antibody pairs (Haab *et al.*, 2001). This group also printed 114 different antibodies on poly-L-lysine-coated slides. Slides were probed with fluorescently labelled antigens in a range of known concentrations and binding was detected and quantified. Interestingly, only 20% of antibody–antigen pairs displayed a linear response in fluorescence with increasing antigen concentration. They speculated that labelling of the antigen might be inefficient as some antigens do not have readily accessible amine groups for labelling with NHS-activated dyes. It was also noted that background fluorescence increased concomitantly with elevated protein concentration, which reduced the precision of quantification of bound protein. They concluded that antibody arrays could detect their target protein at concentrations below $1\,\mathrm{ng\,ml^{-1}}$. However, the optimal total protein concentration should be $<1\,\mathrm{mg\,ml^{-1}}$. The complexity of the protein mixture could perhaps be reduced by a prefractionation step (Haab *et al.*, 2001).

MacBeath and Schreiber (2000) illustrated that it was possible to detect phosphorylation of immobilized proteins whereby three kinase substrates [kemptide, protein phosphatase inhibitor 2 (I-2) and Elk-1] were spotted onto BSA-NHS-coated slides and incubated with a different kinase and [γ^{33}P]ATP. When adenosine 3′,5′-monophosphate-dependent protein kinase was incubated with the slide and the slide developed in photographic emulsion, silver grains were observed where the substrate 'kemptide' had been spotted. Similarly, casein kinase II and p42 mitogen-activated protein (MAP) kinase were shown to phosphorylate I-2 and Elk1 respectively.

Other PTMs can also be readily analysed. For example, proteins can be modified whereby reducing sugars react with N-termini and side chains of amino acids yielding advanced glycosylation end-products. The relative reactivities of immobilized dipeptides were assessed by exposure of a dipeptide spot library to ^{14}C glucose or fructose and autoradiography (Münch *et al.*, 1999).

4.3.3 Applications of proteome arrays

Although the construction of a true proteome array incorporating every ORF expressed by a species or a cell type is an enormously challenging task, several groups have worked towards this goal. Liquid expression cultures and purified His_6-tagged proteins from a human fetal brain library have been arrayed onto PVDF microfilters and readily detected using anti-His_6 antibodies and horseradish peroxidase-conjugated secondary antibodies (Lueking *et al.*, 1999). Such microarrayed libraries could therefore be probed with a chosen ligand and the identity of the arrayed elements to which it binds may be ascertained by returning to the expression cultures and sequencing of cloned insert DNAs (Lueking *et al.*, 1999).

As mentioned previously, Zhu *et al.* (2000) expressed and purified 5800 His_6-tagged yeast proteins and arrayed them, in duplicate, on a nickel-coated glass slide. First, this array was probed with calmodulin to identify yeast proteins that may be regulated by calcium. The array was incubated with biotinylated calmodulin in the presence of calcium and bound calmodulin was detected using Cy3-labelled streptavidin (Zhu *et al.*, 2001). Using this approach, the workers identified 6 known and 33 novel calmodulin-binding proteins. These findings also lead to the identification of a consensus motif present in 14 of these proteins.

Phosphoinositides (PIs) are phospholipids that are present at discrete membrane sites and that function as second messengers in signal transduction pathways (Odorizzi *et al.*, 2000). As the role of PIs in yeast membrane trafficking has not been fully characterized, the yeast proteome array was also screened for PI-interacting proteins (Zhu *et al.*, 2001). Five biotinylated PI liposomes were generated each containing phosphatidylcholine and one of five PIs; including PI(3)P, PI(4)P, PI(3,4)P_2, PI(4,5)P_2 or PI(3,4,5)P_3. Arrays were probed with each liposome and interactions were again detected with Cy3-streptavidin. This screen generated 150 positive signals including a number of membrane-associated proteins and several proteins which appeared to be involved in glucose metabolism (Zhu *et al.*, 2001). Such an array could also be screened for PTMs, enzyme or DNA-binding activity.

Another application of proteome arrays would be to assist in the understanding of the pathogenicity of disease. For instance, a proteome array of a pathogen could be studied for PTMs, interactions with host proteins and enzymatic activity. Moreover, serum could be taken from individuals previously exposed to the pathogen and host antibodies labelled and incubated with the array. Positive signals could indicate which pathogen proteins had elicited the immune response with possible consequences for vaccine studies. This type of array would also be useful for looking at possible drug targets. *Figure 4.5* illustrates the potential applications of proteome arrays.

4.3.4 Applications of custom protein arrays

As with the proteome arrays, custom arrays carrying a defined subset of proteins or variants of a single protein (*Figure 4.6*), face similar challenges to produce. However, several pioneering studies have been performed to assess the capacity and sensitivity of such arrays.

The universal protein array of Ge (2000) carrying immobilized general transcription factors, activators and co-activators was probed with a radiolabelled

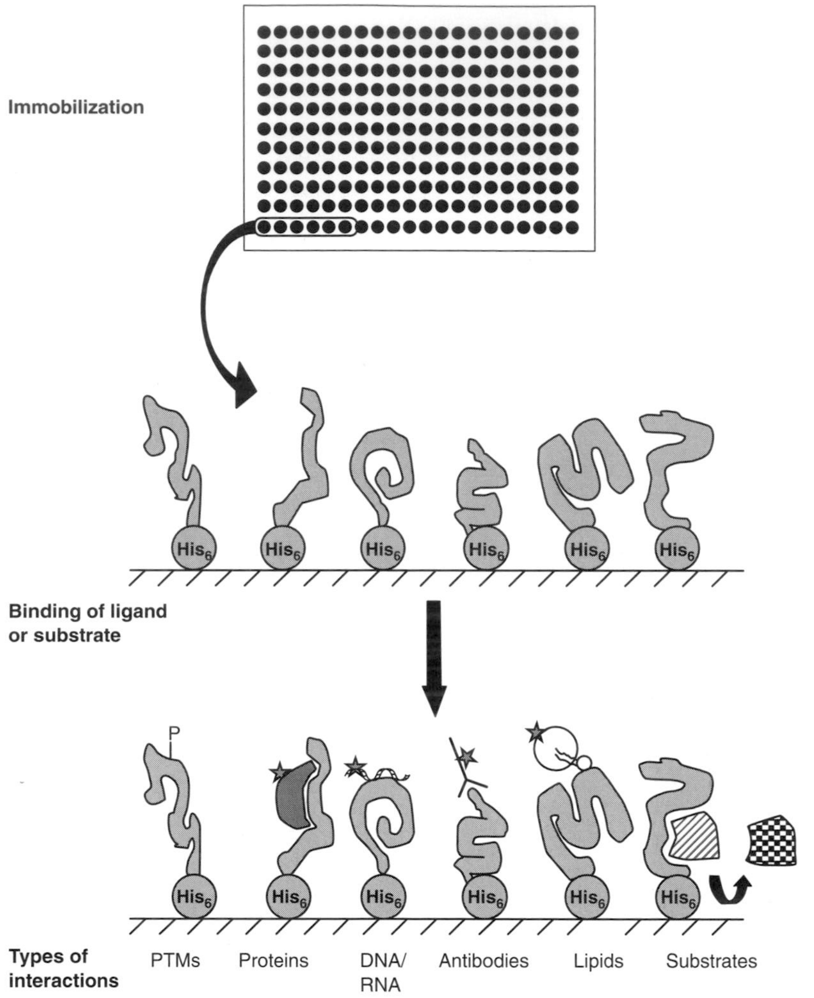

Figure 4.5

Proteome microarrays. ORFs from a specific organism are expressed and the cognate proteins are immobilized on a derivatized solid support in an array format. Unoccupied reactive groups on the support may be quenched using reagents such as BSA or glycine. A single ligand, a mixture of ligands or a substrate may then be applied to the array, which may be subsequently washed to remove non-specifically bound ligand. Binding events may be detected by observing the presence of a label * (e.g. fluorescence or radioactivity), enzyme product, or by the application of a specific labelled antibody. This type of array may be used to detect PTMs e.g. phosphorylation, interactions between proteins, proteins and DNA/RNA, proteins and antibodies, proteins and lipids and enzymatic activities.

novel human transcriptional co-activator p52, and densitometry was used to quantify interactions with p52 binding partners. The same array was also probed with labelled DNA and RNA probes to detect protein–DNA and protein–RNA interactions. A single amino acid change was introduced into a transcriptional co-activator and both wild-type and mutant proteins were

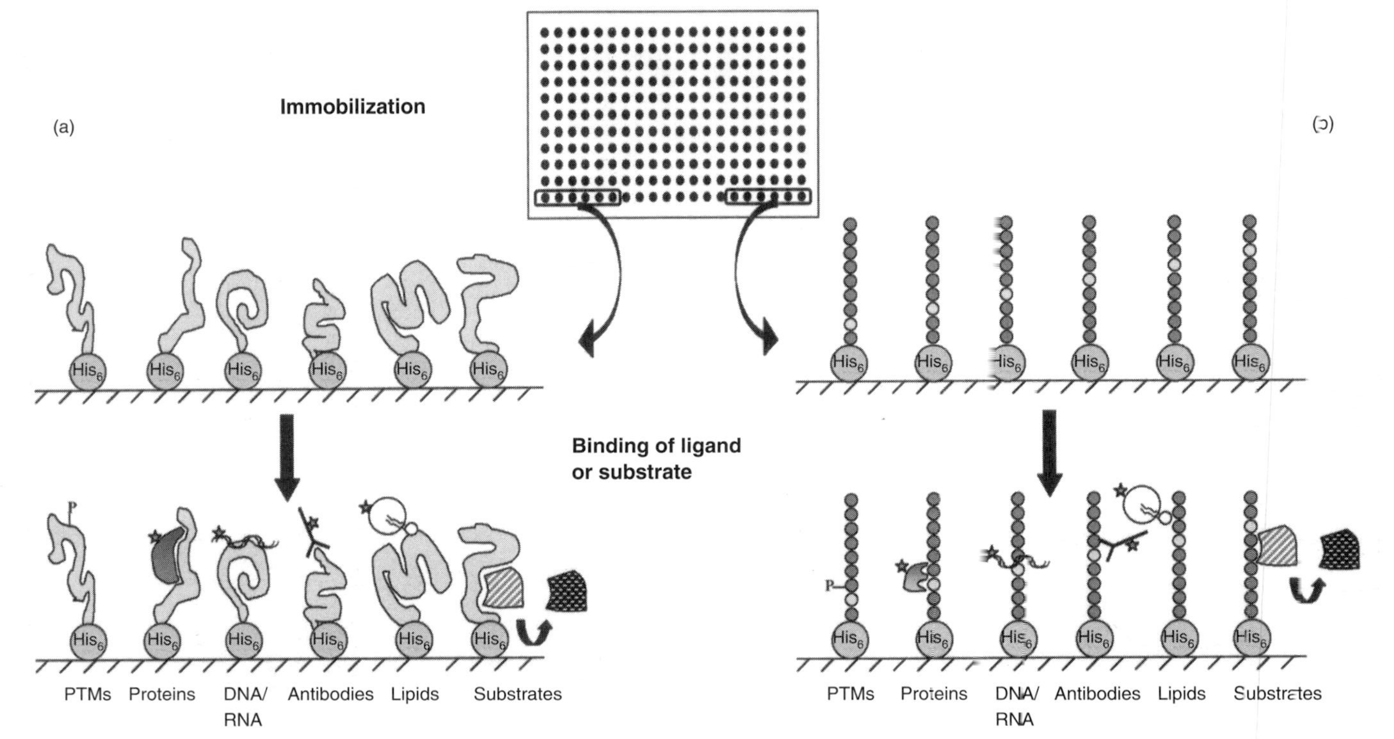

Figure 4.6

Custom protein microarrays. Such arrays consist of a subset of related proteins (a), for example transcription factors, kinases or allergens. Alternatively, they may contain the same protein or peptide bearing a selection of residue changes (b). In each case, the purified components are immobilized on a derivatized solid support in an array format. Unoccupied reactive groups on the support surface may be then quenched using reagents such as BSA or glycine. A single ligand, a mixture of ligands or a substrate is then applied to the array, which may be washed to remove non-specifically bound ligand. Binding events may be detected by observing the presence of a label * (e.g. fluorescence or radioactivity), enzyme product or by the application of a specific labelled antibody. This type of array may be used to detect PTMs, for example, phosphorylation, interactions between proteins, proteins and DNA/RNA, proteins and antibodies, proteins and lipids and enzymatic activities. Furthermore, site-directed mutagenesis of peptides (b) may allow the characterization of binding motifs, epitope mapping and the identification of mutants with altered substrate specificities.

spotted on the UPA. When incubated with DNA and RNA probes, whilst both proteins could bind RNA, the amino acid change completely abolished the ability of the protein to bind dsDNA. This study indicates that it is feasible to perform site-directed mutagenesis and express the products in an array format to assess binding affinities in parallel.

Zhu *et al.* (2000) generated a custom protein microarray onto which 119 yeast kinases were immobilized. Kinases were expressed in yeast as GST-fusion proteins, purified and arrayed on a microchip in nanowells. Initially, [γ^{33}P]ATP was added to the arrayed kinases to look for autophosphorylation events. Subsequently, 17 different substrates and [γ^{33}P]ATP were added individually to each well and phosphorylation was detected using a phosphorimager. Of the 119 kinases, 112 exhibited activity of fivefold or more over background with at least one substrate (Zhu *et al.*, 2000). This study yielded significant amounts of data, including the identification of substrates for 18 uncharacterized kinases, and kinases that are capable of phosphorylating tyrosine residues. Zhu *et al.* (2000), do however acknowledge the fact that activity *in vitro* does not necessarily reflect the situation *in vivo*. Such studies complement MS-based affinity tag approaches to characterize the components of protein complexes that many kinases are associated with (Ho *et al.*, 2002).

Peptide sequences from the same protein may be printed or synthesized on custom arrays and used to identify immunogenic epitopes. Epitopes may be generated by the synthesis of overlapping segments of a protein, immobilization on a solid support and probing with labelled antisera (Geysen *et al.*, 1984; Reineke *et al.*, 2001). Using this technique, Geysen *et al.* (1984) identified a reactive hexapeptide from the coat protein VP1 of the foot-and-mouth virus. This group also synthesized peptides with replacement residues assessing relative reactivities to each variant until the epitope was refined to four specific residues, two of which were essential for antibody binding. This system could be scaled down onto a microchip, whereby multiple copies of the same protein or peptide could be arrayed, each differing at one or more residue to characterize immunogenic epitopes, binding motifs, catalytic domains or sites of PTM (reviewed by Reineke *et al.*, 2001).

A diagnostic application of custom protein microarrays would be to array common antigens to produce an allergy chip, for example. Patient sera may be labelled and exposed to the array to identify which antigens have elicited a powerful immune response. Joos *et al.* (2000) used a similar approach, arraying 18 known autoantigens previously characterized in several autoimmune diseases. Probing of the array with antisera from normal individuals and patients with autoimmune disorders allowed the sensitive detection of autoantibodies titres, thereby providing a cheap and sensitive means of diagnosing autoimmune diseases (Joos *et al.*, 2000).

4.3.5 Applications of protein capture chips

4.3.5.1 Antibody arrays

The use of antibodies or antibody fragments bypasses the requirement for the purification of a vast number of proteins. Antibody arrays can yield information about novel binding partners of a chosen protein involved in a particular biochemical process (*Figure 4.7*). For example, arrays of printed antibodies specific to characterized proteins involved in events such as

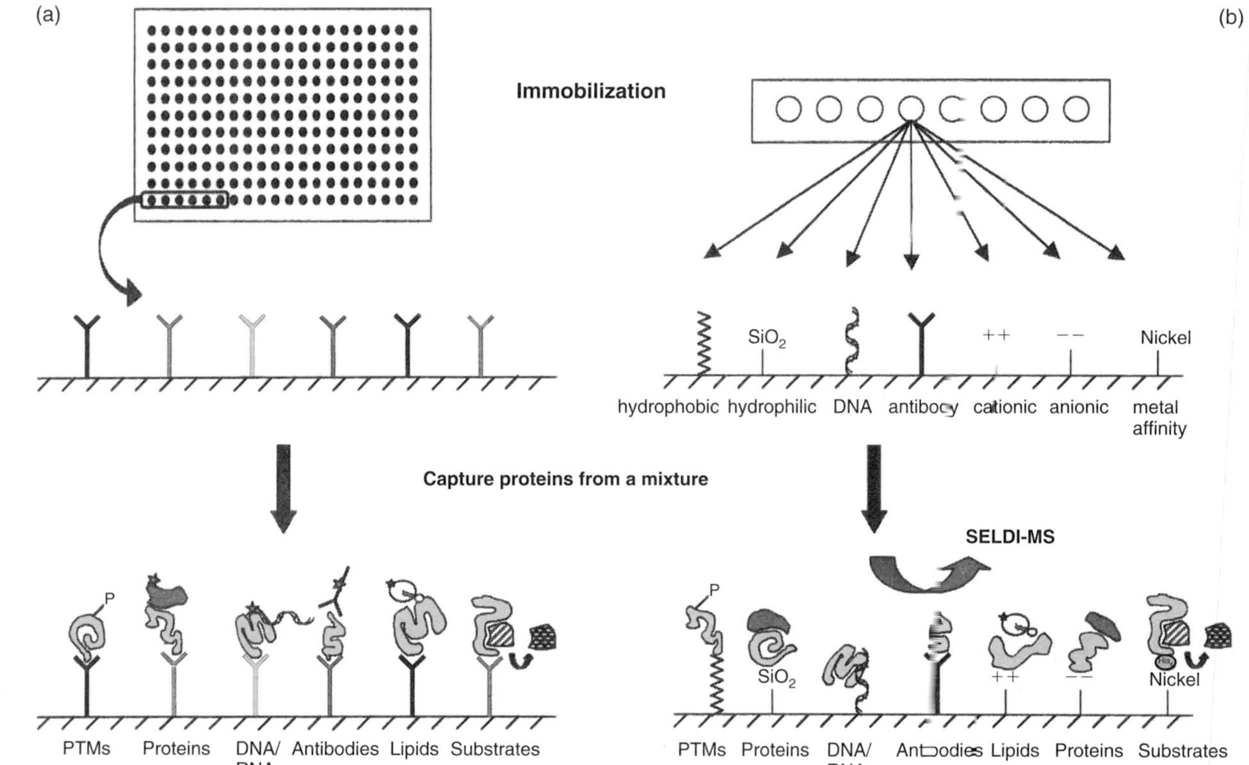

Figure 4.7

Capture microchips. Proteins may be captured from a complex mixture onto an array of immobilized antibodies (a) or onto a chip with up to 16 chemically or biochemically modified surfaces (b), such as the ProteinChip® (Ciphergen Biosystems). In both cases, the protein mixture is applied to the capture chip, which binds proteins specific to the arrayed antibodies or with a particular chemical property, and the array is washed to remove non-specifically bound protein. Binding to the antibody array may be detected by observing the presence of a label * (e.g. fluorescence or radioactivity), or by the application of a labelled antibody specific to the protein of interest. Alternatively, SPR may be used to observe binding and dissociation events. SPR and ProteinChip® technology may be combined with SELDI-MS to characterize bound proteins. This may also involve on-chip digestion with a site-specific protease and peptide mass fingerprinting (reviewed by Merchant & Weinberger, 2000). These types of arrays can be used to detect PTMs, for example, phosphorylation, interactions between proteins, proteins and DNA/RNA, proteins and antibodies, proteins and lipids and enzymatic activities.

apoptosis, the cell cycle or signal transduction are commercially available (e.g. Hypromatrix). Such arrays could be probed with lysates from different cell types or tissue that has been subjected to a range of treatments and the array would recognize and capture a host of different proteins involved in the chosen biochemical process. The washed array could be subsequently interrogated with a labelled antibody specific to the protein of interest. A positive signal would indicate predicted or novel interactions between the captured proteins from the lysate and the protein of interest. When antibodies to the target protein are not available, the protein of interest can be expressed fused to a tag that would be recognized by appropriate antibodies. This type of array may also be used to detect PTMs of proteins, for example, tyrosine phosphorylation of cell cycle-regulating proteins derived from normal or tumour tissue (Hypromatrix).

Antibody arrays have also been utilized to quantify protein expression. Huang (2001), spotted antibodies reactive to cytokines on a membrane and incubated it with sera and media collected from human mammary epithelial cells and human breast cancer cells. The membranes were probed with one of six biotin-labelled antibodies reactive to each cytokine and ECL was used to detect and quantify the level of cytokine present in each sample (Huang, 2001). This type of diagnostic analysis is not only cheap but is highly sensitive with low background noise and could be extended to a microchip format. An alternative approach would be to probe an antibody array with fluorescently labelled samples such as body fluids or tissue lysates. In fact, two tissue lysates, one from normal tissue and one from diseased tissue may be differentially labelled with two fluorophores, mixed and used to probe the array. The ratio between each fluorescence signal can be used to quantify and compare the relative abundance of a range of proteins. Such information could identify up- or down-regulated proteins in diseased versus normal tissue. Novel disease markers may also be identified, which has implications for diagnostic tests.

Capture chips can also, in principle, be combined with SPR and MS. In such cases, a single capture agent such as an antibody or protein is immobilized on a gold chip and a protein solution is passed over the surface. The kinetics of binding and dissociation events is monitored by the biosensor in real-time. This technique has also been described as 'ligand fishing' (reviewed by Nelson *et al.*, 2000). SPR can be combined with MS analysis using proteolytic digestion to identify proteins fished out by the capture agent. For example, the antibody to interleukin (IL)-1α was immobilized on a sensor chip and a protein mixture containing IL-1α was passed over the chip surface (Nelson, 2000). SPR was used to detect the retention of protein on the chip and the captured protein was eluted from the chip and directed to a flow cell containing immobilized pepsin. Following digestion, peptides were subjected to MALDI-MS, which yielded masses corresponding to the IL-1α peptides. This type of microchip can be used to identify unknown ligands bound to an immobilized protein of choice. Furthermore, the approach can harness lipids and membrane-associated proteins since the development of hydrophobic and lipophilic sensor chips (Biacore). SPR techniques do have limitations, however, such as optimization of flow rates, low specificity of proteolytic enzymes due to high or low pHs required for elution, and the requirement to regenerate the immobilized protein after the passage of each analyte.

4.3.5.2 ProteinChips®

ProteinChip® and SELDI-MS technology have powerful applications, particularly in diagnostic research. A complex mixture of proteins from cells or body fluids can be applied to a ProteinChip®, which selectively captures up to several nanomoles of protein exhibiting specific physical properties (*Figure 4.7*) (Merchant & Weinberger, 2000). Once MS has identified a peak of interest, the captured protein can be enriched and subjected to on-chip digestion to enable peptide fingerprinting and accurate identification. This technology may be used to generate protein profiles of particular samples and to identify possible markers of disease (Fung *et al.*, 2001). For example, Wright *et al.* (1999) used ProteinChips® and SELDI-MS to observe peaks corresponding to four prostate cancer-associated protein biomarkers using normal and cancerous epithelial cell lysates, serum and seminal plasma specimens (Wright *et al.*, 1999).

In another ground breaking example, Davies *et al.* (1999), utilized ProteinChip® technology to investigate the role of isoforms of amyloid β peptide (Aβ), a proteolytic fragment derived from amyloid precursor protein that is involved in the pathogenesis of Alzheimer's disease. An anti-amyloid Aβ peptide antibody was immobilized on the ProteinChip® enabling the capture of multiple variants of Aβ secreted into HEK cell media. SELDI-MS was subsequently employed to identify the Aβ variants. This system was also used to observe the profile of isoforms secreted by cells treated with increasing concentrations of cholesterol (Frears *et al.*, 1999).

4.4 Concluding remarks

This chapter has described three types of microchip: proteome arrays, custom protein arrays and protein capture chips. Such systems, particularly in miniaturized form, are recent tools, but their applications are vast. Technological advancement of microarrays will focus on further miniaturization, automation, reduction in sample volumes and improvements in specificity, sensitivity and throughput. Additional challenges that may be addressed include: optimization of membrane protein processing, faster purification of proteins, detection of low affinity interactions and improved mass accuracy. Future advances may move toward solution arrays, whereby peptides may be synthesized or captured on beads (Lam *et al.*, 1991; Fulton *et al.*, 1997) or nanoparticles that harbour a unique identity tag in the form of a fluorophore (Han *et al.*, 2001) or barcode (Zhou *et al.*, 2001). Such solution arrays have advantages over solid-phase arrays in that the capture element need not compromise affinities for its ligand due to inappropriate orientation or the absence of available cofactors, for example. Furthermore, the number of capture elements is not limited to the dimensions of the chip, therefore a vast number of elements may be interrogated by many ligands simultaneously.

Scientific goals that would revolutionize diagnostic and biochemical research would be the production and commercial availability of proteome arrays for every microbial pathogen, for instance, and ultimately arrays of every protein expressed by a species. Such systems would enable scientists and physicians to monitor the abundance and activities of thousands of proteins simultaneously from any cell, thereby allowing the quantitative measurement of protein expression and the generation of predictive models of key cellular processes.

Acknowledgements

Research in the authors' lab is supported by the BBSRC, DEFRA, MRC and the Wellcome Trust.

References

Anderson L, Seilhamer J (1997) A comparison of selected mRNA and protein abundances in human liver. *Electrophoresis* **18**: 533–537.

Arenkov P *et al.* (2000) Protein microchips: use for immunoassay and enzymatic reactions. *Anal Biochem* **278**: 123–131.

Avseenko NV *et al.* (2001) Immobilization of proteins in immunochemical microarrays fabricated by electrospray deposition. *Anal Chem* **73** (24): 6047–6052.

Borrebaeck CA (2000) Antibodies in diagnostics – from immunoassays to protein chips. *Immunol Today* **21**: 379–382.

Borrebaeck CA *et al.* (2001) Protein chips based on recombinant antibody fragments: a highly sensitive approach as detected by mass spectrometry. *Biotechniques* **30**: 1126–1130, 1132.

Bowtell DD (1999) Options available – from start to finish – for obtaining expression data by microarray. *Nat Genet* **21**: 25–32.

Büssow K *et al.* (1998) A method for global protein expression and antibody screening on high-density filters of an arrayed cDNA library. *Nucleic Acids Res* **26**: 5007–5008.

Büssow K *et al.* (2000) A human cDNA library for high-throughput protein expression screening. *Genomics* **65**: 1–8.

Clackson T *et al.* (1991) Making antibody fragments using phage display libraries. *Nature* **352**: 624–628.

Davies H *et al.* (1999) Profiling of amyloid beta peptide variants using SELDI Protein Chip arrays. *Biotechniques* **27**: 1258–1261.

de Wildt RM *et al.* (2000) Antibody arrays for high-throughput screening of antibody–antigen interactions. *Nat Biotechnol* **18**: 989–994.

Eisen MB, Brown PO (1999) DNA arrays for analysis of gene expression. *Methods Enzymol* **303**: 179–205.

Ekins RP (1998) Ligand assays: from electrophoresis to miniaturized microarrays. *Clin Chem* **44**: 2015–2030.

Ekström S *et al.* (2001) Signal amplification using 'spot-on-a-chip' technology for the identification of proteins via MALDI-TOF MS. *Anal Chem* **73**: 214–219.

Emili AQ, Cagney G (2000) Large-scale functional analysis using peptide or protein arrays. *Nat Biotechnol* **18**: 393–397.

Fodor SP *et al.* (1991) Light-directed, spatially addressable parallel chemical synthesis. *Science* **251**: 767–773.

Frank R (1992) Spot-synthesis: an easy technique for the positionally addressable, parallel chemical synthesis on a membrane support. *Tetrahedron* **48**: 9217–9232.

Frears ER *et al.* (1999) The role of cholesterol in the biosynthesis of beta-amyloid. *Neuroreport* **10**: 1699–1705.

Fulton RJ *et al.* (1997) Advanced multiplexed analysis with the FlowMetrix system. *Clin Chem* **43**: 1749–1756.

Fung ET *et al.* (2001) Protein biochips for differential profiling. *Curr Opin Biotechnol* **12**: 65–69.

Ge H (2000) UPA, a universal protein array system for quantitative detection of protein–protein, protein–DNA, protein–RNA and protein–ligand interactions. *Nucleic Acids Res* **28**: e3.

Gershoni JM, Palade GE (1983) Protein blotting: principles and applications. *Anal Biochem* **131**: 1–15.

Geysen HM *et al.* (1984) Use of peptide synthesis to probe viral antigens for epitopes to a resolution of a single amino acid. *Proc Natl Acad Sci USA* **81**: 3998–4002.

Grayhack EJ, Phizicky EM (2001) Genomic analysis of biochemical function. *Curr Opin Chem Biol* **5**: 34–39.

Guschin D *et al.* (1997) Manual manufacturing of oligonucleotide, DNA, and protein microchips. *Anal Biochem* **250**: 203–211.

Gygi SP *et al.* (1999) Correlation between protein and mRNA abundance in yeast. *Mol Cell Biol* **19**: 1720–1730.

Haab BB (2001) Advances in protein microarray technology for protein expression and interaction profiling. *Curr Opin Drug Discov Devel* **4**: 116–123.

Haab BB *et al.* (2001) Protein microarrays for highly parallel detection and quantitation of specific proteins and antibodies in complex solutions. *Genome Biol* **2**: research 0004.1–0004.13

Han M *et al.* (2001) Quantum-dot-tagged microbeads for multiplexed optical coding of biomolecules. *Nat Biotechnol* **19**: 631–635.

Haupt K, Mosbach K (1998) Plastic antibodies: developments and applications. *Trends Biotechnol* **16**: 468–475.

Ho Y *et al.* (2002) Systematic identification of protein complexes in *Saccharomyces cerevisiae* by mass spectrometry. *Nature* **415**: 180–183.

Houseman, BT *et al.* (2002) Peptide chips for the quantitative evaluation of protein kinase activity. *Nat Biotech* **20**: 270–274.

Huang RP (2001) Detection of multiple proteins in an antibody-based protein microarray system. *J Immunol Methods* **255**: 1–13.

Jones VW *et al.* (1998) Microminiaturized immunoassays using atomic force microscopy and compositionally patterned antigen arrays. *Anal Chem* **70**: 1233–1241.

Joos TO *et al.* (2000) A microarray enzyme-linked immunosorbent assay for autoimmune diagnostics. *Electrophoresis* **21**: 2641–2650.

Kanno S *et al.* (2000) Assembling of engineered IgG-binding protein on gold surface for highly oriented antibody immobilization. *J Biotechnol* **76**: 207–214.

Köhler G, Milstein C (1975) Continuous cultures of fused cells secreting antibody of predefined specificity. *Nature* **256**: 495–497.

Kolkman JA, Stemmer WP (2001) Directed evolution of proteins by exon shuffling. *Nature Biotechnol* **19**: 423–428.

Lam KS *et al.* (1991) A new type of synthetic peptide library for identifying ligand-binding activity. *Nature* **354**: 82–84.

Lueking A *et al.* (1999) Protein microarrays for gene expression and antibody screening. *Anal Biochem* **270**: 103–111.

MacBeath G, Schreiber SL (2000) Printing proteins as microarrays for high-throughput function determination. *Science* **289**: 1760–1763.

Mendoza LG *et al.* (1999) High-throughput microarray-based enzyme-linked immunosorbent assay (ELISA). *Biotechniques* **27**: 776–778, 788.

Merchant M, Weinberger SR (2000) Recent advancements in surface-enhanced laser desorption/ionization-time of flight-mass spectrometry. *Electrophoresis* **21**: 1164–1177.

Mooney JF *et al.* (1996) Patterning of functional antibodies and other proteins by photolithography of silane monolayers. *Proc Natl Acad Sci USA* **93**: 12287–12291.

Morozov VN, Morozova TY (1999) Electrospray deposition as a method for mass fabrication of mono- and multicomponent microarrays of biological and biologically active substances. *Anal Chem* **71**: 3110–3117.

Münch G *et al.* (1999) Amino acid specificity of glycation and protein-AGE cross-linking reactivities determined with a dipeptide SPOT library. *Nat Biotechnol* **17**: 1006–1010.

Nelson RW *et al.* (2000) Biosensor chip mass spectrometry: a chip-based proteomics approach. *Electrophoresis* **21**: 1155–1163.

Nocito A *et al.* (2001) Tissue microarrays (TMAs) for high-throughput molecular pathology research. *Int J Cancer* **94**: 1–5.

Odorizzi G *et al.* (2000) Phosphoinositide signaling and the regulation of membrane trafficking in yeast. *Trends Biochem Sci* **25**: 229–235.

Owens RM *et al.* (2001) Copurification of the Lac repressor with polyhistidine-tagged proteins in immobilized metal affinity chromatography. *Protein Expr Purif* **21**: 352–360.

Reineke U *et al.* (2001) Applications of peptide arrays prepared by the SPOT-technology. *Curr Opin Biotechnol* **12**: 59–64.

Roberts RW, Szostak JW (1997) RNA–peptide fusions for the *in vitro* selection of peptides and proteins. *Proc Natl Acad Sci USA* **94**: 12297–12302.

Rogers YH *et al.* (1999) Immobilization of oligonucleotides onto a glass support via disulfide bonds: a method for preparation of DNA microarrays. *Anal Biochem* **266**: 23–30.

Silzel JW *et al.* (1998) Mass-sensing, multianalyte microarray immunoassay with imaging detection. *Clin Chem* **44**: 2036–2043.

Stemmer WP (1994) Rapid evolution of a protein *in vitro* by DNA shuffling. *Nature* **370**: 389–391.

Timofeev E *et al.* (1996) Regioselective immobilization of short oligonucleotides to acrylic copolymer gels. *Nucleic Acids Res* **24**: 3142–3148.

Tovey ER, Baldo BA (1989) Protein binding to nitrocellulose, nylon and PVDF membranes in immunoassays and electroblotting. *J Biochem Biophys Methods* **19**: 169–184.

Vasiliskov AV *et al.* (1999) Fabrication of microarray of gel-immobilized compounds on a chip by copolymerization. *Biotechniques* **27**: 592–598, 600.

Walter G *et al.* (2000) Protein arrays for gene expression and molecular interaction screening. *Curr Opin Microbiol* **3**: 298–302.

Wenschuh H *et al.* (2000) Coherent membrane supports for parallel microsynthesis and screening of bioactive peptides. *Biopolymers* **55**: 188–206.

Williams C (2000) Biotechnology match making: screening orphan ligands and receptors. *Curr Opin Biotechnol* **11**: 42–46.

Wright Jr GL *et al.* (1999) ProteinChip® surface enhanced laser desorption/ionization (SELDI) mass spectrometry: a novel protein biochip technology for detection of prostate cancer biomarkers in complex protein mixtures. *Prostate Cancer and Prostatic Diseases* **2**: 264–276.

Xia Y *et al.* (1996) Complex optical surfaces formed by replica molding against elastomeric masters. *Science* **273**: 347–349.

Zhou H *et al.* (2001) Solution and chip arrays in protein profiling. *Trends Biotechnol* **19** (10, Suppl): S34–S39.

Zhu H *et al.* (2000) Analysis of yeast protein kinases using protein chips. *Nat Genet* **26**: 283–289.

Zhu H *et al.* (2001) Global analysis of protein activities using proteome chips. *Science* **293**: 2101–2105.

Zhu H, Snyder M (2001) Protein arrays and microarrays. *Curr Opin Chem Biol* **5**: 40–45.

Single Nucleotide Polymorphism Genotyping Using Microarrays

5

Tomi Pastinen

5.1 Introduction

The unprecedented advances in human genome sequencing have generated an extensive catalogue of commonly occurring DNA variants within the human population, most of which are single nucleotide polymorphisms (SNPs). The publicly and privately funded SNP discovery efforts are all directed at facilitating the anticipated mapping and characterization of genetic risk factors for common diseases and providing a genetic basis for the observed interindividual variations in drug response. Genetic risk profiles for common diseases, elucidation of novel pathogenic mechanisms and tailored drug therapies are the supposed outcomes of future genetic studies. Regardless of the study design, the use of SNP resources requires a very high number of data points (or genotypes) per sample in the discovery phase and a moderate number of genotypes per sample in the 'diagnostic' or targeted phase that possibly follows. For example, estimates of the required number of genotypes per sample in whole genome association studies vary from tens to hundreds of thousands, and even with the development of the 'human haplotype map' (Daly *et al.*, 2001) the numbers can be as high as 10^5 SNPs per sample (Patil *et al.*, 2001). Clearly, such studies are of prohibitive size with current genotyping throughput and cost.

This chapter reviews the advances and promise of DNA microarray technology for high-throughput genotyping and targeted resequencing. Other applications of the assay format are described elsewhere in this book. Much of the technology utilized is common among microarrays and the emphasis of this chapter is on solutions applied to genotyping.

5.2 DNA microarrays for sequence analysis

5.2.1 Background

A series of theoretical papers and patent applications published 15 years ago by several independent groups introduced the sequencing by hybridization (SBH) approach (Drmanac & Crkvenjakov, 1987; Bains & Smith, 1988; Lysov *et al.*, 1988; Southern, 1988; Khrapko *et al.*, 1989; Bains, 1991). The

Microarrays & Microplates: Applications in Biomedical Sciences, Shu Ye and Ian N.M. Day

original SBH technology was intended for *de novo* sequencing by hybridization, which was believed to have potential for very high throughput and to be easily automated.

The simple idea of reading a sequence based on the hybridization reaction onto its constituent DNA-oligomers was presented in two formats. Format I had target DNA immobilized on a solid support followed by sequential queries using labelled hybridization probes (Strezoska *et al.*, 1991; Drmanac *et al.*, 1993). Format II had a large number of oligonucleotide probes immobilized either on polyacrylamide gel pads (Khrapko *et al.*, 1989, 1991) or synthesized directly on to a derivatized glass surface (Southern *et al.*, 1992) followed by hybridization of a labelled target.

De novo sequence analysis was complicated by short oligonucleotide probes (usually octamers), as it was not feasible to construct *n*-mer arrays with sufficient probe length because of the unrealistic number of different oligonucleotides required (Southern, 1996), for example, a complete set of 15-mers would require 10^9 probes. The problem of low yield hybridization of AT-rich probes was recognized at an early stage of SBH trials (Southern *et al.*, 1992). 'Stacking' hybridization (Broude *et al.*, 1994; Yershov *et al.*, 1996; Stomakhin *et al.*, 2000) was suggested to improve the SBH results. But interest has shifted from *de novo* sequencing to multiplex genotyping and comparative sequencing.

5.2.2 Array production and readout

Sequence analysis on microarrays is based on oligonucleotide probes interrogating the amplified DNA target. Oligonucleotide arrays are constructed by deposition and immobilization of different oligomers in spatially addressable sites on a 2D surface with a high density. If a certain continuous DNA fragment is to be scanned for sequence variation, a tiled array design is employed which contains overlapping sets of oligonucleotides designed to interrogate successive base pairs in the target sequence. Monitoring recurrent variation at several different targets or sites within the same target is usually carried out using probe sets interrogating only these sites of interest. The third approach is to manufacture all possible sequences of a given length (*n*-mer arrays) on to an array.

5.2.2.1 *In situ* synthesis

The array can be manufactured either by combinatorial *in situ* synthesis or pre-made probe deposition on a derivatized surface, sometimes referred to as 'off-chip' manufacture. *In situ* synthesis by standard phosphoramidite chemistry on derivatized glass surfaces was pioneered by Southern and colleagues (Southern *et al.*, 1992; Maskos and Southern, 1992, 1993a). Their synthesis method produced tiled arrays with all possible probe lengths along the synthesis cell path (Southern *et al.*, 1994). Another *in situ* approach utilized standard phosphoramidites on a derivatized poly-propylene surface delivered by a multi-channel fluidic system (Matson *et al.*, 1994, 1995). Neither of the above approaches have been applied to large-scale genotyping. A more recent system utilized *in situ* synthesis with standard phosphoramidites on surface tension-treated glass to confine the synthesis to defined areas (Butler *et al.*, 2001).

Fodor and colleagues (1991) introduced a sophisticated method for the synthesis of biopolymers on planar surfaces. The method combined semiconductor-based photolithography and solid-phase chemical synthesis to achieve highly parallel *in situ* synthesis of biopolymers on small glass surfaces. Using phosphoramidites with photolabile 5′-protective groups, Affymetrix demonstrated the synthesis of 256 different octanucleotides on a 1.28 cm^2 surface in just 16 chemical coupling steps (Pease *et al.*, 1994) – DNA-arrays were now called 'DNA-chips'. The drawback of producing high-density arrays using the Affymetrix method has been the low step-wise yield of synthesis, varying from 92 to 94%, effectively limiting the probe lengths to 20–25 bp (McGall *et al.*, 1997) and alternative approaches have been explored to improve the synthesis yields (McGall *et al.*, 1996; Wallraff *et al.*, 1997).

5.2.2.2 Spotted arrays

The 'off-chip' synthesis of oligonucleotides and their deposition in minute volumes with various methods on different activated surfaces provides a more accessible method for DNA-array manufacture. Photopolymerized gel pads can be produced in relatively simple steps and oligonucleotides can be covalently immobilized to these pads (Khrapko *et al.*, 1991; Yershov *et al.*, 1996; Guschin *et al.*, 1997; Proudnikov *et al.*, 1998; Vasiliskov *et al.*, 1999). Glass surfaces can be derivatized using various chemical approaches (Lindroos *et al.*, 2001) to allow covalent attachment of oligonucleotides via amino- or disulfide groups. Recently, chemistries creating dendrimeric structures on glass surfaces increasing the surface area binding the oligonucleotides (Beier & Hoheisel 1999; Benters *et al.*, 2002) and immobilization of acrylamide-modified oligonucleotides via co-polymerization (Rehman *et al.*, 1999) have been presented.

Robotic devices based on contact printing pins (Schena *et al.*, 1995; Shalon *et al.*, 1996), inkjet dispensing heads (Lemmo *et al.*, 1998; Stimpson *et al.*, 1998, Okamoto *et al.*, 2000), nanolitre dispensing needles (Graves *et al.*, 1998) and electrospray deposition (Morozov & Morozova, 1999) can be applied to deliver minute droplets of DNA probes on to surfaces, achieving densities of up to 10^5 probes per cm^2. Special array designs utilize electronic addressing of charged DNA probes to affinity capture sites (Sosnowski *et al.*, 1997), selective polymerization of acrylamide on optical fibre tips (Healey *et al.*, 1997) or randomly ordered, labelled microspheres binding DNA probes on optic fibres (Steemers *et al.*, 2000).

5.2.2.3 Established array readout systems

Early applications of DNA arrays invariably used ^{32}P-labelled probes and phosphorimaging detection systems. Radioisotopic labelling with ^{33}P coupled with phosphorimaging detection provides slightly better spatial resolution (i.e. Southern *et al.*, 1994; Pastinen *et al.*, 1997, 1998; Drmanac *et al.*, 1998; Mir and Southern, 1999), but is still limiting compared with the array synthesis densities. Development of fluorescence labelling schemes and detection systems based on an epifluorescence confocal scanner with photomultiplier tube (PMT) detector (Fodor *et al.*, 1991; Pease *et al.*, 1994; Lipshutz *et al.*, 1995) or a fluorescence microscope coupled to a cooled CCD camera (Mirzabekov 1994; Yershov *et al.*, 1996) led to improved resolution. There are now several commercial providers of confocal array scanners, supplied with two to five excitation sources suitable for fluorophores absorbing at 488–650 nm and emitting at 515–690 nm. Also CCD-based systems with white light excitation are commercially available.

5.2.2.4 Experimental array detection systems

In addition to phosphorimaging and confocal laser scanning systems, several other detection approaches of array read-out have been suggested. Direct integration of the DNA array with a CCD (Eggers *et al.*, 1994) or phototransistor sensing element (Vo-Dinh *et al.*, 1999) offers promise of sensitive detection of fluorescent signals in a highly compact format. Arrays formed by bundles of optic fibres with probes attached directly to their distal ends (Healey *et al.*, 1997) or on coded beads attached to the fibres (Steemers *et al.*, 2000) have been described. Illumina (San Diego, CA) has recently commercialized the coded bead array approach (BeadArray™), but no large-scale applications have been published on the platform thus far. Evanescent wave excitation and total internal reflection have been applied to microarray systems (Stimpson *et al.*, 1995). Recently, an instrument with four excitation lasers utilizing the evanescent wave principle to detect four-colour primer extension signals was presented (Kurg *et al.*, 2000). Gold nanoparticles coupled with scanometric detection have high sensitivity (Taton *et al.*, 2000), and a similar approach with electrical detection not only improved the sensitivity compared with traditional detection systems, but also allowed very high mismatch discrimination in a model experiment (Park *et al.*, 2002). Quantum dot nanocrystals offer attractive spectral features for multi-colour applications, and could allow high levels of multiplexing with spectrally distinct labels (Han *et al.*, 2001). The use of an entirely different detection system on solid gold surfaces based on differential charge transduction through matched vs. mismatched DNA duplexes by cyclic voltammometry has recently been presented (Kelley *et al.*, 1999).

5.3 Arrays for resequencing and SNP-scoring

Comparative sequencing in a broad sense includes both resequencing of known sequence with unknown SNPs and scoring of known polymorphic bases. The resequencing and SNP-scoring applications are discussed separately below as they differ significantly in the experimental approaches applied to them.

5.3.1 Resequencing arrays

5.3.1.1 Template preparation for resequencing

For resequencing long contiguous stretches of DNA are analysed (few hundred base pairs to several kilobases), so generally the templates are prepared by PCR or long-range PCR without multiplexing. In hybridization-based systems the amplified targets are then fragmented and labelled by haptens or fluorophores to improve hybridization kinetics.

5.3.1.2 High-density (Affymetrix) resequencing arrays

In complete sequence analysis 'tiled' sets of probes immobilized on the microarray are used to interrogate each successive base. Such tiled arrays require a large number of probes (the standard is 8 × the number of base pairs interrogated). High-density light-directed synthesized arrays from Affymetrix (Santa Clara, CA) have dominated as primary platforms for resequencing arrays, as it has been the only format able to accommodate the required number probes.

The earliest versions of such arrays were designed to interrogate cystic fibrosis transmembrane conductance regulator gene (*CFTR*) exon 11 sequence with a minimal set of tiled 15-mer probes each synthesized on 365 × 365 μm synthesis sites (Cronin *et al.*, 1996). At this point it was already stated that the simplest form of scanning array (four probes per interrogated nucleotide) would not be sufficiently sensitive to detect heterozygous mutations. Regions of the HIV-1 genome known to be associated with resistance to antiviral therapy were assayed on the Affymetrix chips (Kozal *et al.*, 1996), this resequencing array had a considerably more complex design (32 probes per interrogated nucleotide) and was shown to be equal in accuracy to Sanger dideoxy-sequencing. An independent evaluation of the commercial Affymetrix HIV-1 chip demonstrated that mutations presenting 50% of the studied viral population could not be detected reliably (Gunthard *et al.*, 1998), however.

Chee and colleagues (1996) applied the Affymetrix chips to a considerably larger target – the entire mitochondrial genome – using over 130 000 different probes to interrogate the sequence. A two-colour labelling strategy to include an internal control was employed. A good genotyping result with 98–99% accuracy in base calling was achieved in this haploid genome. The first large human genomic application of the high-density arrays was resequencing of an exon of the *BRCA1* gene (Hacia *et al.*, 1996), in which 14 of 15 tested patient samples were scored correctly.

In the first large SNP survey more than 10^9 oligos were synthesized on 149 chips, covering a 2 Mb stretch of genomic DNA, the sensitivity and specificity were both reported to be below 90% in this study (Wang *et al.*, 1998). Similarly, a sequence survey of the *ATM* gene (Hacia *et al.*, 1998a) and evaluation of the commercial *p53* gene array (Ahrendt *et al.*, 1999) resulted in sensitivities of 88–91%. Resequencing for SNP discovery consequently used denaturing high-performance liquid chromatography (DHPLC) in parallel with high-density arrays to achieve higher sensitivity and specificity (Cargill *et al.*, 1999); alternatively, SNPs detected on the array were treated as 'candidate SNPs' (Halushka *et al.*, 1999). The discovery of mouse SNP using high-density arrays was reported to be more successful (Lindblad-Toh *et al.*, 2000), possibly because of the use of inbred homozygous mouse strains.

The most recent resequencing applications have chosen different strategies to tackle the problem of imperfect detection of heterozygous sites in large-scale applications (i.e. a false-positive rate of up to 45% and a 30% false-negative rate). The application of rigorous statistical algorithms to the Affymetrix-chip data yielded 80% success and perfect specificity in calling over 8×10^5 diploid sites (Cutler *et al.*, 2001). Interestingly, in the larger chromosome 21 resequencing effort a much larger proportion of the data (65% success) was discarded to yield a 3% miscall rate in haploid sites (Patil *et al.*, 2001). The apparent discrepancy in success rate among these recent studies is explained by the 15% failure in template preparation (PCR) in the latter study. The discrepancy in specificity, 0% miscalls in diploid sites vs. 3% miscalls in haploid sites, remains unexplained. Based on these sequence surveys on Affymetrix high-density arrays, one can conservatively estimate that the technology currently provides up to 80% sensitivity and >95% specificity.

5.3.1.3 Approaches to improve resequencing array performance

In all resequencing applications on high-density arrays, a proportion of sites yields consistently low signal intensities, some of these poorly hybridizing probes have been recovered using 5-methyluridine triphosphates in the target (Hacia *et al.*, 1998b). Affymetrix has also sought to increase the flexibility of their sequence scanning arrays by applying a ligation reaction on generic, partially double-stranded 8- or 9-mer arrays to detect sequence differences between a test and a control sequence (Gunderson *et al.*, 1998). The performance of the 9-mer arrays was excellent with targets up to 1.2 kb. Similarly Head *et al.* (1997) utilized the fidelity of a DNA polymerase in model experiments scanning a 33-bp stretch of the *p53* gene on primer extension arrays. An intriguing approach would be to use polymerase extension on high-density arrays, which might be possible with alternative array synthesis procedures (McGall *et al.*, 1996; Wallraff *et al.*, 1997) or a novel inversion strategy for primers attached at their 3′-end (Kwiatkowski *et al.*, 1999). It was recently shown that the discrimination of SNPs on hybridization-based arrays suffers from truncated probes (Jobs *et al.*, 2002), which are generated during the light-directed *in situ* synthesis of arrays.

The less explored SBH format involving the immobilization of targets on filters and successive interrogation with short oligonucleotide probes in solution has been shown to be effective in determining sequence variation in stretches of cloned DNA (Drmanac *et al.*, 1998). The use of this strategy is complicated by the need for several thousand hybridization reactions to deduce the sequence, and is thus limited to large groups automating the successive hybridization steps (Drmanac & Drmanac, 1999).

5.3.2 Scoring SNPs or mutations on DNA microarrays

Genotyping of previously characterized SNPs at several different loci by DNA microarrays has different key requirements to resequencing. Preparation of the template to enrich the several genomic fragments spanning the variants of interest is more complex because of the large number of targets, and the need for virtually 100% specificity in allele scoring is challenging. Furthermore, the growing number of known SNP public and private databases has created a high demand for massively parallel SNP scoring. Lastly, in the allele scoring phase, the number of samples to analyse ranges from hundreds to thousands in typical genomic studies. *Figure 5.1* shows the commonly employed approaches for allele discrimination on microarrays.

5.3.2.1 Target preparation for SNP scoring by PCR

Amplification of the target DNA and reduction of its complexity are achieved by PCR. Parallel analysis of SNPs or mutations necessitates the use of multiplex PCR, for which several procedures have been presented (e.g. Chamberlain & Chamberlain, 1994; Shuber *et al.*, 1995; Henegariu *et al.*, 1997; Zangenberg *et al.*, 1999). In practice, each application requires separate optimization of the multiplex PCR reaction if the aim is to achieve a 100% success rate in amplification (Hacia *et al.*, 1998a; Cheng *et al.*, 1999; Pastinen *et al.*, 2000). Generic protocols to avoid optimization of individual reactions, to allow a more efficient assay set-up and to increase levels of multiplexing have been applied with tolerance to losses of amplifiable genomic fragments

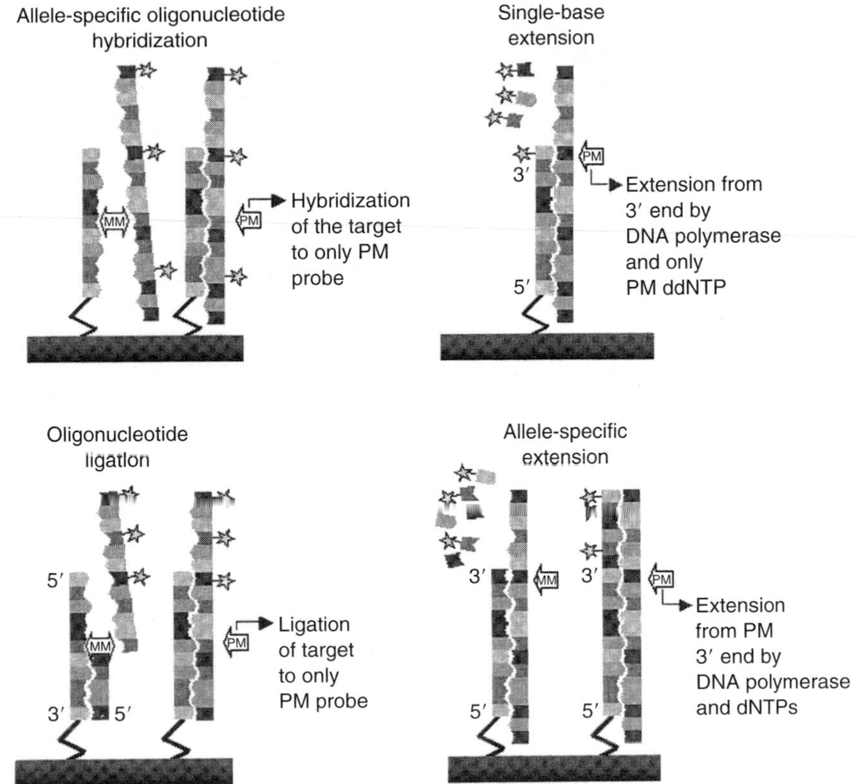

Figure 5.1

Methods for allelic discrimination on DNA microarrays. In allele-specific oligohybridization each site of interest is queried by two or more immobilized oligonucleotides. The labelled target (marked with stars) binds preferentially to the perfectly matched (PM) probe, whereas hybridization to the mismatched probe (MM) is thermodynamically less stable. Single base extension is based on the fidelity of DNA polymerase to incorporate only the nucleotide complementary (PM) to the site of interest. The immobilized oligonucleotide is designed to anneal just 5′ to the site of interest and the extension can be carried out either in parallel reactions for each of the four nucleotides or differentially labelled dideoxynucleotides can be used in a single extension reaction. Solid-phase ligation on microarrays has only been presented for resequencing using partially double-stranded probes immobilized on the surface (Gunderson *et al.*, 1998). The targets are fragmented and labelled, only the 5′-ends complementary to the probes are ligated by a ligase and remain bound to immobilized probes following higher stringency washes. In allele-specific extension two probes with 3′-ends complementary to one or the other allele are immobilized. The immobilized probes are extended by a DNA polymerase using labelled dNTPs, the enzyme only extends the probe with 3′-end complementary to the site of interest.

(Wang *et al.*, 1998; Cho *et al.*, 1999). The different array platforms allow parallel analysis of hundreds or thousands of polynucleotides, whereas at present multiplex PCRs can only be extended to tens of fragments in parallel. Thus, in allele scoring the array size is not limiting compared with resequencing arrays, and this allows use of the various available technologies for low-to-medium density array manufacture. For hybridization-based systems, the amplified targets can be labelled with a second round of PCR using labelled nucleosides followed by the fragmentation of labelled targets. In primer extension the labelling actually takes place directly on the array, but the targets still need to be fragmented or rendered single-stranded prior to array analysis.

5.3.2.2 Experimental ASO hybridization-based methods

In two short reports Southern and Maskos (1993a, b) first described the optimization of ASO probes for the detection of three beta-globin alleles, followed by the synthesis of optimal probes on a solid surface and the genotyping of four samples. Similarly, Yershov *et al.* (1996) suggested hybridization for mutation screening on oligonucleotides immobilized to gel pads using the beta-globin gene as a model. 'Stacking' hybridization probes and two-colour hybridization was used to study three variable sites. Alternatively, melting curves of the hybrids were measured in real time to improve allelic discrimination, which was assessed using five amplified targets and several synthetic targets on arrays for five different nucleotide positions (Drobyshev *et al.*, 1997). Various modifications of ASO hybridization-based microarray genotyping have been presented. Guo *et al.* (1994) used glass supports and determined that spacer length, surface density and use of a single-stranded target were important for good hybridization yields. Another study did not find spacer arms separating the ASO probes from the glass surface critical, but only three samples were tested for two mutations (Beattie *et al.*, 1995). *In situ* synthesized peptide nucleic acid (PNA) probes were evaluated in model systems indicating some difficulties in predicting behavior of DNA–PNA hybrids and thus limiting their usefulness in genotyping for the time being (Weiler *et al.*, 1997). Electronically addressable electrode array hybridization was first studied using model systems for DNA and PNA probes (Edman *et al.*, 1997; Sosnowski *et al.*, 1997). The assay was applied for genotyping three mannose-binding protein (MBP) gene SNPs and one IL-1b gene SNP (Gilles *et al.*, 1999). Despite the claimed advantages of the electronically addressable arrays and the use of 'electronic stringency' it remains questionable whether these arrays will become popular for routine genotyping as they are complex to manufacture and assay procedures require dedicated instruments. Similar constraints apply to the use of the randomly ordered fibreoptic arrays with probes immobilized to coded beads (Steemers *et al.*, 2000). Piezoelectric *in situ* synthesized microarrays were recently optimized for detection of seven common polymorphisms in the *NAT2* gene (Cronin *et al.*, 2001), this study exemplified the need of individual probe optimization to achieve robust allele calling by ASO-based microarrays. In highly polymorphic genetic systems such as the HLA genes microarrays with redundant ASO probes containing more than one mismatch per probe can yield good genotyping results (Guo *et al.*, 2002).

5.3.2.3 High-density arrays for SNP scoring

The above variants of ASO hybridization on DNA microarrays provided proof-of-principle for the approach, but have not been implemented in 'standard' SNP scoring. High-density arrays generated by light-directed synthesis have been applied in larger scale studies. Thirty-two *CFTR* alleles were interrogated on DNA arrays to determine the genotypes of 10 blinded samples (Cronin *et al.*, 1996). The first SNP-scoring application presented by Wang and colleagues (1998) applied 46-plex PCRs to amplify genomic targets, which were pooled for hybridization analysis. Fewer than 75% of the >500 sites performed well enough to allow genotype assignment. One quarter of the successful SNPs were validated in three individuals and two Centre d'Etude du Polymorphisme Humain (CEPH) families with a good success rate (98%) and high confidence assignment of genotypes (99.9%). The same arrays were applied by Hacia *et al.* (1999) to determine allele frequencies at 214 markers using pooled samples from different populations. An identical approach was used to genotype *Arabidopsis thaliana* SNPs (Cho *et al.*, 1999). Almost half of the markers had to be discarded in this study because of imperfect discrimination of genotypes on the *Arabidopsis* arrays. These studies indicate that there will be a considerable number of polymorphisms that are not amenable to high-density array scoring: at present it is unclear whether it will be possible to predict which sites will be difficult to genotype. Furthermore, high-level PCR multiplexing is 'costly' as 10–20% of markers are lost at this stage.

A commercial array (HuSNP™, Affymetrix, Santa Clara, CA) with probe sets to detect 1494 different SNPs is claimed to yield 1200–1300 genotypes per sample, translating into a success rate of 80–87% (Genechip® HuSNP™ Mapping Assay, Technical Note No. 1, Part No. 700318, Affymetrix), an example of an allele call from this array is shown in *Figure 5.2*. The HuSNP™ array has been applied in two published studies, one examining loss-of-heterozygosity in tumour tissue-derived samples (Lindblad-Toh *et al.*, 2000b) and another investigating SNPs associated with substance abuse in pooled samples (Uhl *et al.*, 2001). The former study yielded an average 80% (1205/1494) call rate on SNPs per array assay and a miscall rate of 4.6%, whereas the latter did not report these values. Importantly, the SNPs scored on the HuSNP™ array were originally discovered using the same platform (Wang *et al.*, 1998), thus a miscall rate approaching 5% is quite high. Furthermore, the arrays for SNP scoring have higher redundancy than the resequencing arrays, which suggests that the development of accurate hybridization-based

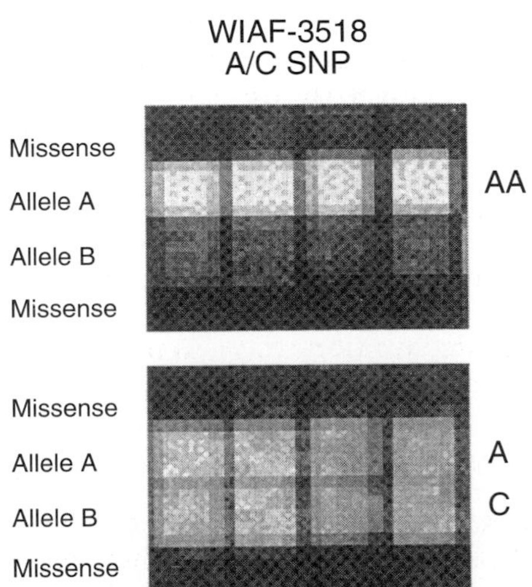

Figure 5.2

SNP scoring by hybridization on high-density microarrays. Probe sets for the two alleles (A and B) are composed of 25-mers with the SNP complementary to base position 9, 12, 13 or 14 and additional missense probes are generated by light-directed *in situ* synthesis. The example shown is derived from the commercial HuSNP™ array comprising similar probe sets for over 1400 SNPs. The upper probe set shows intense signals only for probes corresponding to allele A and the sample is called a homozygote (AA), whereas the lower probe set shows equal signals from probes corresponding to both alleles and the sample is called a heterozygote (CA). (Image courtesy of Dr K. Lindblad-Toh, Whitehead Institute/MIT Center for Genome Research, Cambridge, MA).

allele scoring systems is challenging. It is clear that higher specificity (i.e. a lower miscall rate) is required for large-scale genetic studies.

An approach to avoid the requirement of specific PCR amplification for each SNP analysed was recently suggested by Affymetrix (Dong *et al.*, 2001). The 'generic' amplification strategy was based on restriction enzyme digestion of the whole human genome followed by adapter ligation, generic adapter specific PCR, size selection and resequencing (on microarrays) for SNP discovery. The discovered SNPs were scored on the same resequencing arrays after enriching the polymorphic targets using biotinylated fragment-specific probes coupled with affinity capture and the universal amplification strategy. Multiplexed amplification of 92 targets was shown to be possible, 77 of which were true SNPs (16% false-positive rate in array-based SNP discovery).

5.3.2.4 DNA-modifying enzymes in microarray genotyping

'Traditional' reaction formats have shown that the use of DNA polymerases and ligases can improve genotype discrimination under uniform reaction conditions, thus multiplexed genotyping should be more feasible with the aid of these enzymes. The solid-phase reactions on microarrays have thus far applied mainly DNA polymerases and primer extension, whereas solid-phase ligation systems have not been scrutinized in detail.

In mini-sequencing on DNA microarrays detection primers are immobilized at their 5′-end and designed to anneal just 5′ to the nucleotide on the target; DNA polymerase is then used to incorporate labelled ddNTPs complementary to the site of interest with high specificity (*Figure 5.1*; Pastinen *et al.*, 1997). The same reaction principle has been denoted arrayed primer extension (APEX; Shumaker *et al.*, 1996; Kurg *et al.*, 2000), nested GBA (Head *et al.*, 1997) and 'multibase single stranded primer extension' (Dubiley *et al.*, 1999). The optimized primer extension system was shown to have an order of magnitude higher allelic discrimination on DNA microarrays than ASO hybridization on the same platform (Pastinen *et al.*, 1997). Fluorescent single nucleotide primer extension was applied for analysis of 10 allelic variants of the beta-globin gene (Kurg *et al.*, 2000). The primer extension was performed in both orientations of the template with a set of four ddNTPs each labelled with a different fluorophore and readout was based on a four-colour evanescent wave laser excitation system. The average genotype discrimination was nearly 40-fold in the nine tested samples. Larger numbers of samples by primer extension on microarrays have been analysed in two published studies, one examining 12 candidate gene SNPs for coronary heart disease in a case–control population of 300 individuals (Pastinen *et al.*, 1998) and another studying 25 Y-chromosomal SNPs in 320 individuals (Raitio *et al.*, 2001). In both of these larger studies the allele discrimination was high and ranged from 4- to >100-fold; these results suggest that enzyme-assisted array methods should provide lower miscall rates than hybridization arrays. A recent large-scale application scoring over 1200 SNPs in chromosome 22 using the APEX primer extension arrays (*Figure 5.3*) shows the potential of the approach for scoring complex samples in a highly parallel manner (Remm & Metspalu, 2002; Dawson *et al.*, manuscript submitted).

Another polymerase-assisted primer extension assay is based on two immobilized detection primers with 3′-end complementary to one or the other allele, denoted as 'multiprimer extension assay' (Dubiley *et al.*, 1999) or

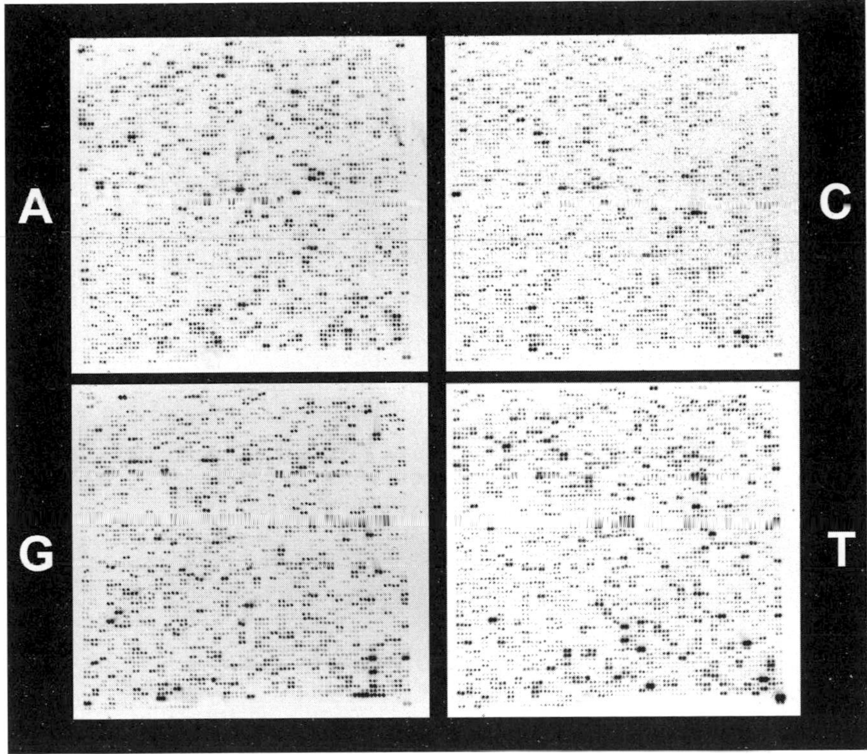

Figure 5.3

Large-scale SNP scoring by single-base extension (APEX). An array for scoring over 1200 SNPs in human chromosome 22 using a four-colour primer extension system (arrayed primer extension, APEX) is shown. The arrays are synthesized by spotting oligonucleotides using a robotic 'pin-and-loop' spotter (GMS 417, Affymetrix) and the solid-phase primer extension results are carried out using four dideoxynucleotides labelled with spectrally resolvable dyes, each imaging channel is shown separately (A, C, G, T). Up to 90% of genotypes could be called using the system (Remm & Metspalu, 2002). The array was applied among other genotyping methods to characterize LD-patterns in chromosome 22, although only 655 of the 1279 SNPs on the array had high enough allele frequency to be included in the LD-analysis (Dawson et al., manuscript submitted). (Image courtesy of Drs Andres Metspalu and Ants Kurg, Estonian Biocentre, Tartu, Estonia.)

'allele-specific extension assay' (Pastinen *et al.*, 2000). A pairwise comparison of a total of 56 genotype calls at seven sites using the single base extension and the multi-primer extension assay yielded similar allele discrimination (Dubiley *et al.*, 1999). The array-based allele-specific extension assay was applied in a large carrier screening study, in which over 2200 samples were analysed (Pastinen *et al.*, 2001), in which all assays were carried out in duplicate (>4500 reactions on arrays), 95% of assays could be called and all the blinded carrier samples were identified correctly. In total over 60 000 array-based genotyping calls were made in this study. Analysis of the largest sample population in any array genotyping study thus far was facilitated by the use of 'an-array-of-arrays' reaction format (see *Figure 5.5*), which enabled analysis of 80 samples per microscope glass slide (Pastinen *et al.*, 2000).

5.3.2.5 Enzyme-based assays coupled with generic microarrays

A molecular barcoding strategy was originally presented for monitoring the phenotypes of yeast deletion strains (Shoemaker *et al.*, 1996). A generic barcode or tag-array is composed of a random, non-cross-hybridizing set of oligonucleotides, which are used to capture multiplexed reaction products tagged with their complement, thus the tag-arrays serve as ways to decode the multiplexed reaction products. The advantage of the approach is the generic array design, which is suitable for any set of SNPs. Furthermore, the genotyping reactions are performed in solution potentially yielding higher success than solid-phase reactions.

Ligation with tagged probes (*Figure 5.4*) has thus far been applied for screening known mutations on generic DNA-microarrays (Gerry *et al.*, 1999; Favis *et al.*, 2000). Arrays with <10 'zip-code' oligonucleotides were

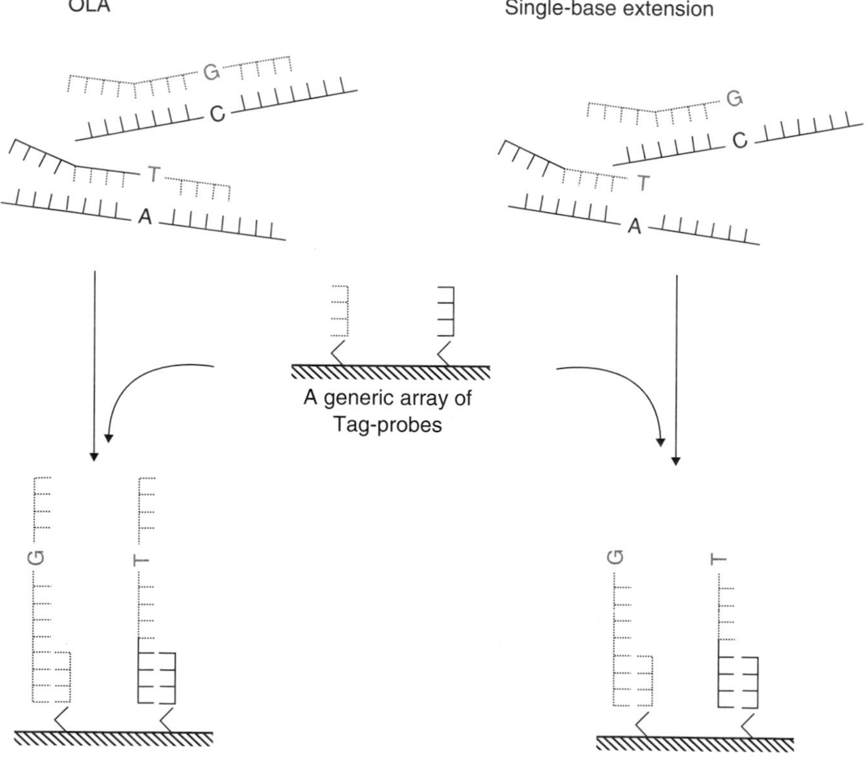

Figure 5.4

Principles for genotyping on generic 'tag-arrays'. Primer extension or oligonucleotide ligation reaction (OLA) is carried out in solution using probes with 5′ sequences (non-aligning tails of the G and T probes in the figure) complementary (cTag) to array immobilized Tag-sequences. Each SNP queried in the solution reaction is coded with a different 5′ cTag sequence. Optimally, the tag-sequences are minimally cross-hybridizing and possess similar hybridization behaviour (melting temperature). The multiplexed reaction products are hybridized on the tag-arrays serving to decode the results. In principle any solution reaction with oligonucleotides is amenable for tag-array-based decoding; an example of an allele-specific extension reaction read out on tag-arrays is shown in *Figure 5.5*.

produced and allele-specific ligation probes with 5′ sequence complementary to one of the oligonucleotides immobilized on the array were used in the ligation reaction. Following ligation in solution, the ligated products were hybridized on the zip-code arrays, and mutated alleles in the K-*ras*, *BRCA1* and *BRCA2* genes could be detected easily. The multiplexing capacity of the ligation assay was limited by the detection sensitivity, which could be improved by signal amplification using rolling circle amplification and 'decorator' probe labelling (Ladner *et al.*, 2001).

Larger panels of SNPs have been screened on using the tagged single base pair extension method (TAG-SBE) (*Figure 5.4*). The method is based on a generic array of tag-sequences and multiplex SBE reaction performed in solution using differentially labelled ddNTPs with each detection primer carrying a unique 5′ sequence complementary to one of the tag-sequences immobilized on the array. Following the SBE reaction, the primers are hybridized on the tagged arrays and the incorporated nucleotides are identified by a multi-colour array reader. Hybridization of the tagged products can be carried out both on Affymetrix high density arrays (Fan *et al.*, 2000) and also on custom-spotted tag arrays (Hirschhorn *et al.*, 2000). Both studies reported successful genotyping at over 100 SNPs, with 20–35% of sites failing due to low signal intensities or ambiguous genotype clusters; at successful sites >95% of genotypes could be called and validation yielded >99% accuracy of the called genotypes. The reactions (PCR and SBE) were multiplexed up to 50-fold and the tag-array hybridizations were multiplexed up to 140-fold.

We recently applied custom-made generic arrays for decoding multiplexed allele-specific primer extension products (*Figures 5.5* and *5.6*). Both PCR amplification and the allele-specific extension reaction are multiplexed up to 48-fold and the 'array-of-arrays' accommodates up to 80 separate hybridization reactions on a standard microscopic glass slide. Further multiplexing in the hybridization phase is achieved by using four different fluorophores in the extension reactions, allowing analysis of four individuals in each hybridization reaction, the maximum capacity of the system is ≈30 000 SNPs per standard microscopic glass slide. A general procedure for carrying out SNP scoring with this method is described in the Protocols section. The section is divided into eight steps: primer design, array preparation and spotting, multiplex PCR, post-PCR processing (*in vitro* transcription), primer extension reaction in solution, tag-array hybridization and result interpretation.

5.4 Summary and future directions

Hybridization-based SNP resequencing systems have been applied at sufficiently large scales to assess their performance in the discovery of new SNPs and mutations. A single hybridization reaction on high-density resequencing arrays can provide good quality sequence for up to 80% of base pairs in a ≈50 kb target sequence. The error frequency of the method is approaching that of the Sanger dideoxysequencing method at least for haploid genomes. Current applications of the resequencing arrays are large-scale SNP discovery and dedicated diagnostic assays. Apart from the commercialized diagnostic assays the system is still being applied only by a handful of centres. It is of note that, despite the remarkable advances in this technology over the past

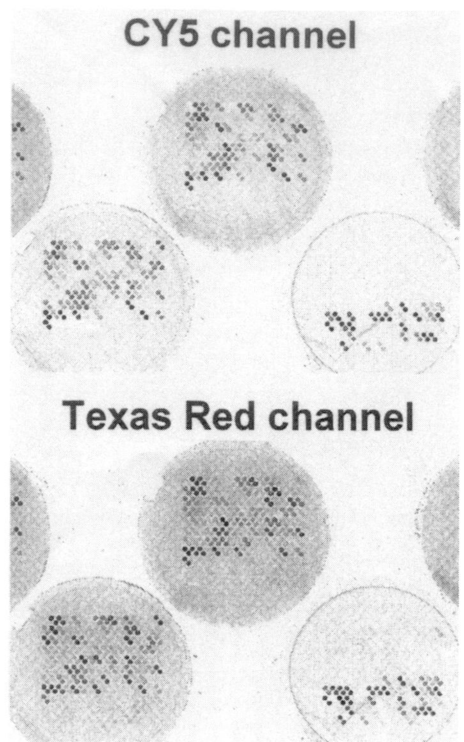

Figure 5.5

Multiplexed allele-specific extension coupled with tag-array detection. We applied the tag-arrays for visualizing multiplexed allele-specific primer extension products. Each SNP is scored using two allele-specific probes differing by their 3′ base and the cTag-sequence at the 5′ end of the probe; the reactions are carried out under conditions similar to those described previously for solid-phase extension (Pastinen *et al.*, 2000). The 'array-of-arrays' format is employed to tag-array hybridizations of large sample sizes. Three replicate arrays are shown in the figure, a separate hybridization solution is applied to each with a custom-made rubber grid separating the samples during hybridization. In the example shown each replicate array contains 192 probes in duplicate (spotted by a piezoelectric dispenser), enabling analysis of 96 SNPs per array and each slide contains 40–80 separable arrays. The second feature increasing the throughput of our system is the use of multiple dyes in extension reaction, each of which corresponds to a different sample analysed on the same array. Fourfold multiplexing is easily achievable with different fluorescent dyes (2/4 channels shown). Lastly, the flexibility of the tag-array system is exemplified in the figure, which shows two arrays (lower left corner and middle) hybridized with reaction products of 48-plex SNP-scoring products, whereas the third replicate array (lower right corner) is used to visualize a separate 24-plex reaction. Importantly, some of the same tags are used in the two completely different applications; this generic nature of the tag-arrays are their major strength. An example of the allele calling from the 48-plex SNP-scoring application is shown in *Figure 5.6*.

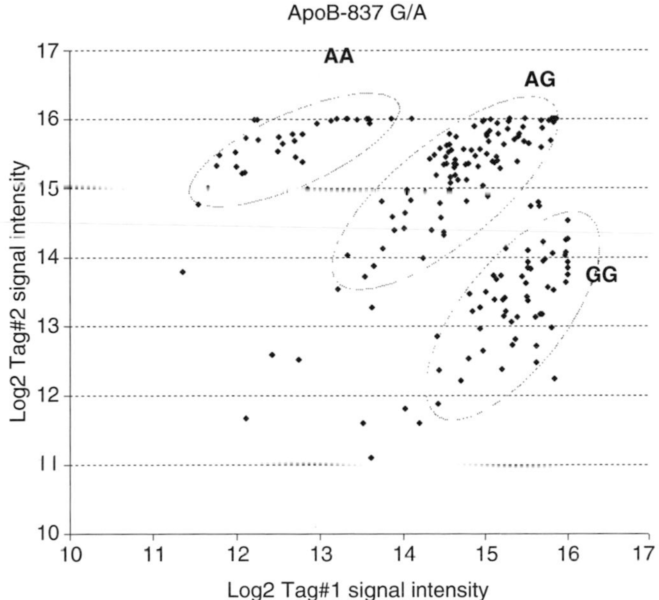

Figure 5.6

Allele-scoring from tag-array hybridization of multiplexed allele-specific extension products. Detection of the APOB-837 G/A SNP in 192 individuals, including 14 three-generation CEPH families, by allele-specific extension coupled with tag-arrays. Tag #1 corresponds to the G-allele (log2 transformed intensity values plotted on *x*-axis) and Tag #2 corresponds to A-allele (*y*-axis) as described in the Protocol section. The results are derived from a four-colour 48-plex PCR/extension assay hybridized to custom-made tag-arrays (*Figure 5.5*), the total number of parallel assays in the shown experiment was 9216 (48 × 192) and 75% of SNPs yielded sufficiently high signal intensities to be clustered. The scoring of the shown SNP yielded 180 genotyping calls (94%, the called samples are circled with the genotype call shown) and there were no Mendelian errors in the called genotypes. The experiments were carried out as detailed in the Protocol section.

10 years, the promise of a fully automated read-out of genotyping results is yet to be fulfilled. Even in the recent large-scale applications the final quality of array calls was based on visual inspection of the array images. This requirement of 'manual' data analysis is likely due to highly variable locus-(sequence) dependent signal intensity level and allele discrimination in these complex hybridization assays. However, no other microarray platform seems currently practical for resequencing applications, as the light-directed combinatorial *in situ* synthesis remains the only way to produce arrays of sufficient complexity.

The current bottleneck in SNP scoring is template amplification rather than achievable array complexity. While the largest SNP scoring systems have been based on hybridization on high-density microarrays, their imperfect genotype discrimination (Lindblad-Toh *et al.*, 2000b) has prompted the development of alternative assays. The enzyme-based allele discrimination assays in either solid-phase or solution-phase reactions show promise for accurate large-scale SNP scoring and first large-scale applications have been published. The maturation of MALDI-TOF-based large-scale genotyping methods (Lechner *et al.*, 2002) poses a considerable pressure to speed this development of a widely applicable and standardized enzyme-based microarray platform. Such development is meaningful, particularly in the context of the anticipated genome-wide SNP-association studies, which seem to overwhelm the capacity of any existing system. With the large number of assays required per sample ($\approx 10^4$ to 10^5) even the amount of valuable sample DNA could become limiting in non- or low-level-multiplexed reactions not to mention the limitations posed by the cost and throughput of such approaches. It will be interesting to see whether highly multiplexed microarray-based SNP-scoring assays will fulfil their promise of 'ultra-high throughput' genotyping in the near future. The building blocks seem to be in place to increase the throughput and cost-efficiency of genotyping by 10–100-fold in the near future compared with other high-throughput genotyping platforms. Next-generation solutions for SNP scoring on microarrays are likely to incorporate some of the novel detection technologies and perhaps alternatives to multiplex PCR in target preparation to fully realize the analysis capacity of microarrays.

References

Ahrendt SA *et al.* (1999) Rapid p53 sequence analysis in primary lung cancer using an oligonucleotide probe array. *Proc Natl Acad Sci USA* **96**: 7382–7387.

Bains W (1991) Hybridization methods for DNA sequencing. *Genomics* **11**: 294–301.

Bains W, Smith GC (1988) A novel method for nucleic acid sequence determination. *J Theor Biol* **135**: 303–307.

Baner J *et al.* (1998) Signal amplification of padlock probes by rolling circle replication. *Nucleic Acids Res* **26**: 5073–5078.

Beattie WG *et al.* (1995) Hybridization of DNA targets to glass-tethered oligonucleotide probes. *Mol Biotechnol* **4**: 213–225.

Beier M, Hoheisel JD (1999) Versatile derivatisation of solid support media for covalent bonding on DNA-microchips. *Nucleic Acids Res* **27**: 1970–1977.

Benters R *et al.* (2002) DNA microarrays with PAMAM dendritic linker systems. *Nucleic Acids Res* **30**: E10.

Broude NE *et al.* (1994) Enhanced DNA sequencing by hybridization. *Proc Natl Acad Sci USA* **91**: 3072–3076.

Butler JH *et al.* (2001) *In situ* synthesis of oligonucleotide arrays by using surface tension. *J Am Chem Soc* **123**: 8887–8894.

Cargill M *et al.* (1999) Characterization of single-nucleotide polymorphisms in coding regions of human genes. *Nat Genet* **22**: 231.

Chamberlain JS *et al.* (1988) Deletion screening of the Duchenne muscular dystrophy locus via multiplex DNA amplification. *Nucleic Acids Res* **16**: 11141–11156.

Chamberlain JS, Chamberlain JR (1994) Optimization of multiplex PCRs. In KB Mullis *et al.*, editors. *Polymerase Chain Reaction*, pp. 38–46. Boston: Birkhäuser.

Chee M *et al.* (1996) Accessing genetic information with high-density DNA arrays. *Science* **274**: 610–614.

Cheng S *et al.* (1999) A multilocus genotyping assay for candidate markers of cardiovascular disease risk. *Genome Res* **9**: 936–949.

Cho RJ *et al.* (1999) Genome-wide mapping with biallelic markers in *Arabidopsis thaliana*. *Nat Genet* **23**: 203–207.

Cronin MT *et al.* (1996) Cystic fibrosis mutation detection by hybridization to light-generated DNA probe arrays. *Hum Mutat* **7**: 244–255.

Cronin MT *et al.* (2001) Utilization of new technologies in drug trials and discovery. *Drug Disp Metab* **29**: 586–591.

Cutler DJ *et al.* (2001) High-throughput variation detection and genotyping using microarrays. *Genome Res* **11**: 1913–1925.

Daly MJ *et al.* (2001) High-resolution haplotype structure in the human genome. *Nat Genet* **29**: 229–232.

Dong S *et al.* (2001) Flexible use of high-density oligonucleotide arrays for single-nucleotide polymorphism discovery and validation. *Genome Res* **11**: 1418–1424.

Drmanac R, Crkvenjakov R (1987) Yugoslav Patent Application 570.

Drmanac R, Drmanac S (1999) cDNA screening by array hybridization. *Methods Enzymol* **303**: 165–178.

Drmanac R *et al.* (1993) DNA sequence determination by hybridization: a strategy for efficient large-scale sequencing. *Science* **260**: 1649–1652.

Drmanac S *et al.* (1998) Accurate sequencing by hybridization for DNA diagnostics and individual genomics. *Nat Biotechnol* **16**: 54–58.

Drobyshev A *et al.* (1997) Sequence analysis by hybridization with oligonucleotide microchip: identification of beta-thalassemia mutations. *Gene* **188**: 45–52.

Dubiley S *et al.* (1999) Polymorphism analysis and gene detection by minisequencing on an array of gel immobilized primers. *Nucleic Acids Res* **27**: e19.

Edman CF *et al.* (1997) Electric field directed nucleic acid hybridization on microchips. *Nucleic Acids Res* **25**: 4907–4914.

Eggers M *et al.* (1994) A microchip for quantitative detection of molecules utilizing luminescent and radioisotope reporter groups. *Biotechniques* **17**: 516–525.

Fan JB *et al.* (2000) Parallel genotyping of human SNPs using generic high-density oligonucleotide tag arrays. *Genome Res* **10**: 853–860.

Favis R *et al.* (2000) Universal DNA array detection of small insertions and deletions in BRCA1 and BRCA2. *Nat Biotechnol* **18**: 561–564.

Fodor SP *et al.* (1991) Light-directed, spatially addressable parallel chemical synthesis. *Science* **251**: 767–773.

Gerry NP *et al.* (1999) Universal DNA microarray method for multiplex detection of low abundance point mutations. *J Mol Biol* **292**: 251–262.

Gilles PN *et al.* (1999) Single nucleotide polymorphic discrimination by an electronic dot blot assay on semiconductor microchips. *Nat Biotechnol* **17**: 365–370.

Graves DJ *et al.* (1998) System for preparing microhybridization arrays on glass slides. *Anal Chem* **70**: 5085–5092.

Gunderson KL *et al.* (1998) Mutation detection by ligation to complete *n*-mer DNA arrays. *Genome Res* **8**: 1142–1153.

Gunthard HF *et al.* (1998) Comparative performance of high-density oligonucleotide sequencing and dideoxynucleotide sequencing of HIV type 1 pol from clinical samples. *AIDS Res Hum Retroviruses* **14**: 869–876.

Guo Z *et al.* (1994) Direct fluorescence analysis of genetic polymorphisms by hybridization with oligonucleotide arrays on glass supports. *Nucleic Acids Res* **22**: 5456–5465.

Guo Z *et al.* (2002) Oligonucleotide arrays for high-throughput SNPs detection in the MHC Class I genes: HLA-B as a model system. *Genome Res* **12**: 447–457.

Guschin D *et al.* (1997) Manual manufacturing of oligonucleotide, DNA, and protein microchips. *Anal Biochem* **250**: 203–211.

Hacia JG *et al.* (1996) Detection of heterozygous mutations in BRCA1 using high density oligonucleotide arrays and two-colour fluorescence analysis. *Nat Genet* **14**: 441–447.

Hacia JG *et al.* (1998a) Strategies for mutational analysis of the large multiexon ATM gene using high-density oligonucleotide arrays. *Genome Res* **8**: 1245–1258.

Hacia JG *et al.* (1998b) Enhanced high density oligonucleotide array-based sequence analysis using modified nucleoside triphosphates. *Nucleic Acids Res* **26**: 4975–4982.

Hacia JG *et al.* (1999) Determination of ancestral alleles for human single-nucleotide polymorphisms using high-density oligonucleotide arrays. *Nat Genet* **22**: 164–167.

Halushka MK *et al.* (1999) Patterns of single-nucleotide polymorphisms in candidate genes regulating blood-pressure homeostasis. *Nat Genet* **22**: 239.

Han M *et al.* (2001) Quantum-dot-tagged microbeads for multiplexed optical coding of biomolecules. *Nat Biotechnol* **19**: 631–635.

Head SR *et al.* (1997) Nested genetic bit analysis (N-GBA) for mutation detection in the p53 tumor suppressor gene. *Nucleic Acids Res* **25**: 5065–5071.

Healey BG *et al.* (1997) Fiberoptic DNA sensor array capable of detecting point mutations. *Anal Biochem* **251**: 270–279.

Henegariu O *et al.* (1997) Multiplex PCR: critical parameters and step-by-step protocol. *Biotechniques* **23**: 504–511.

Hirschhorn JN *et al.* (2000) SBE-TAGS: an array-based method for efficient single-nucleotide polymorphism genotyping. *Proc Natl Acad Sci USA* **97**: 12164–12169.

Jobs M *et al.* (2002) Effect of oligonucleotide truncation on single-nucleotide distinction by solid-phase hybridization. *Anal Chem* **74**: 199–202.

Kelley SO *et al.* (1999) Single-base mismatch detection based on charge transduction through DNA. *Nucleic Acids Res* **27**: 4830–4837.

Khrapko KR *et al.* (1989) An oligonucleotide hybridization approach to DNA sequencing. *FEBS Lett* **256**: 118–122.

Khrapko KR *et al.* (1991) A method for DNA sequencing by hybridization with oligonucleotide matrix. *DNA Sequencing* **1**: 375–388.

Kozal MJ *et al.* (1996) Extensive polymorphisms observed in HIV-1 clade B protease gene using high-density oligonucleotide arrays. *Nat Med* **2**: 753–759.

Kurg A *et al.* (2000) Arrayed primer extension: solid phase four-color DNA resequencing and mutation detection technology. *Genetic Testing* **4**: 1–7.

Kwiatkowski M *et al.* (1999) Inversion of *in situ* synthesized oligonucleotides: improved reagents for hybridization and primer extension in DNA microarrays. *Nucleic Acids Res* **27**: 4710–4714.

Ladner DP *et al.* (2001) Multiplex detection of hotspot mutations by rolling circle-enabled universal microarrays. *Lab Invest* **81**: 1079–1086.

Lechner D *et al.* (2002) Large-scale genotyping by mass spectrometry: experience, advances and obstacles. *Curr Opin Chem Biol* **6**: 31–38.

Lemmo AV *et al.* (1998) Inkjet dispensing technology: applications in drug discovery. *Curr Opin Biotechnol* **9**: 615–617.

Lindblad-Toh K *et al.* (2000a) Large-scale discovery and genotyping of single nucleotide polymorphisms in the mouse. *Nat Genet* **24**: 381–386.

Lindblad-Toh K *et al.* (2000b) Loss-of-heterozygosity analysis of small-cell lung carcinomas using single-nucleotide polymorphism arrays. *Nat Biotechnol* **18**: 1001–1005.

Lindroos K *et al.* (2001) Minisequencing on oligonucleotide microarrays: comparison of immobilisation chemistries. *Nucleic Acids Res* **29**: E69.

Lipshutz RJ *et al.* (1995) Using oligonucleotide probe arrays to access genetic diversity. *Biotechniques* **19**: 442–447.

Lysov P *et al.* (1988) Determination of the nucleotide sequence of DNA using hybridization with oligonucleotides. A new method. *Dokl Akad Nauk SSSR* **303**: 1508–1511.

Maskos U, Southern EM (1992) Oligonucleotide hybridizations on glass supports: a novel linker for oligonucleotide synthesis and hybridization properties of oligonucleotides synthesised *in situ*. *Nucleic Acids Res* **20**: 1679–1684.

Maskos U, Southern EM (1993a) A novel method for the analysis of multiple sequence variants by hybridisation to oligonucleotides. *Nucleic Acids Res* **21**: 2267–2268.

Maskos U, Southern EM (1993b) A novel method for the parallel analysis of multiple mutations in multiple samples. *Nucleic Acids Res* **21**: 2269–2270.

Matson RS *et al.* (1994) Biopolymer synthesis on polypropylene supports. I. Oligonucleotides. *Anal Biochem* **217**: 306–310.

Matson RS *et al.* (1995) Biopolymer synthesis on polypropylene supports: oligonucleotide arrays. *Anal Biochem* **224**: 110–116.

McGall G *et al.* (1996) Light-directed synthesis of high-density oligonucleotide arrays using semiconductor photoresists. *Proc Natl Acad Sci USA* **93**: 13555–13560.

McGall G *et al.* (1997) The efficiency of light-directed synthesis of DNA arrays on glass substrates. *J Am Chem Soc* **119**: 5081.

Mir KU, Southern EM (1999) Determining the influence of structure on hybridization using oligonucleotide arrays. *Nat Biotechnol* **17**: 788–792.

Mirzabekov AD (1994) DNA sequencing by hybridization – a megasequencing method and a diagnostic tool? *Trends Biotechnol* **12**: 27–32.

Morozov VN, Morozova TYa (1999) Electrospray deposition as a method for mass fabrication of mono- and multicomponent microarrays of biological and biologically active substances. *Anal Chem* **71**: 3110–3117.

Okamoto T *et al.* (2000) Microarray fabrication with covalent attachment of DNA using bubble jet technology. *Nat Biotechol* **18**: 438.

Park SJ *et al.* (2002) Array-based electrical detection of DNA with nanoparticle probes. *Science* **295**: 1503–1506.

Pastinen T *et al.* (1997) Minisequencing: a specific tool for DNA analysis and diagnostics on oligonucleotide arrays. *Genome Res* **7**: 606–614.

Pastinen T *et al.* (1998) Array-based multiplex analysis of candidate genes reveals two independent and additive genetic risk factors for myocardial infarction in the Finnish population. *Hum Mol Genet* **7**: 1453–1462.

Pastinen T *et al.* (2000) A system for specific, high-throughput genotyping by allele-specific primer extension on microarrays. *Genome Res* **10**: 1031–1042.

Pastinen T *et al.* (2001) Dissecting a population genome for targeted screening of disease mutations. *Hum Mol Genet* **10**: 2961–2972.

Patil N *et al.* (2001) Blocks of limited haplotype diversity revealed by high-resolution scanning of human chromosome 21. *Science* **294**: 1719–1723.

Pease AC *et al.* (1994) Light-generated oligonucleotide arrays for rapid DNA sequence analysis. *Proc Natl Acad Sci USA* **91**: 5022–5026.

Proudnikov D *et al.* (1998) Immobilization of DNA in polyacrylamide gel for the manufacture of DNA and DNA-oligonucleotide microchips. *Anal Biochem* **259**: 34–41.

Raitio M *et al.* (2001) Y-chromosomal SNPs in Finno-Ugric-speaking populations analyzed by minisequencing on microarrays. *Genome Res* **11**: 471–482.

Rehman FN *et al.* (1999) Immobilization of acrylamide-modified oligonucleotides by co-polymerization. *Nucleic Acids Res* **27**: 649–655.

Remm M, Metspalu A (2002) High-density genotyping and linkage disequilibrium in the human genome using chromosome 22 as a model. *Curr Opin Chem Biol* **6**: 24–30.

Schena M *et al.* (1995) Quantitative monitoring of gene expression patterns with a complementary DNA microarray. *Science* **270**: 467–470.

Shalon D *et al.* (1996) A DNA microarray system for analyzing complex DNA samples using two-color fluorescent probe hybridization. *Genome Res* **6**: 639–645.

Shoemaker DD *et al.* (1996) Quantitative phenotypic analysis of yeast deletion mutants using a highly parallel molecular bar-coding strategy. *Nat Genet* **14**: 450–456.

Shuber AP *et al.* (1995) A simplified procedure for developing multiplex PCRs. *Genome Res* **5**: 488–493.

Shumaker JM *et al.* (1996) Mutation detection by solid phase primer extension. *Hum Mutat* **7**: 346–354.

Sosnowski RG *et al.* (1997) Rapid determination of single base mismatch mutations in DNA hybrids by direct electric field control. *Proc Natl Acad Sci USA* **94**: 1119–1123.

Southern EM (1988) Analyzing polynucleotide sequences. International Patent Application PCT GB 89/00460.

Southern EM (1996) DNA chips: analysing sequence by hybridization to oligonucleotides on a large scale. *Trends Genet* **12**: 110–115.

Southern EM *et al.* (1992) Analyzing and comparing nucleic acid sequences by hybridization to arrays of oligonucleotides: evaluation using experimental models. *Genomics* **13**: 1008–1017.

Southern EM *et al.* (1994) Arrays of complementary oligonucleotides for analysing the hybridisation behaviour of nucleic acids. *Nucleic Acids Res* **22**: 1368–1373.

Steemers FJ *et al.* (2000) Screening unlabelled DNA targets with randomly ordered fiber-optic gene arrays. *Nat Biotechnol* **18**: 91–94.

Stimpson DI *et al.* (1995) Real-time detection of DNA hybridization and melting on oligonucleotide arrays by using optical wave guides. *Proc Natl Acad Sci USA* **92**: 6379–6383.

Stimpson DI *et al.* (1998) Parallel production of oligonucleotide arrays using membranes and reagent jet printing. *Biotechniques* **25**: 886–890.

Stomakhin AA *et al.* (2000) DNA sequence analysis by hybridization with oligonucleotide microchips: MALDI mass spectrometry identification of 5mers contiguously stacked to microchip oligonucleotides. *Nucleic Acids Res* **28**: 1193–1198.

Strezoska Z *et al.* (1991) DNA sequencing by hybridization: 100 bases read by a non-gel-based method. *Proc Natl Acad Sci USA* **88**: 10089–10093.

Taton TA *et al.* (2000) Scanometric DNA array detection with nanoparticle probes. *Science* **289**: 1757–1760.

Uhl GR *et al.* (2001) Polysubstance abuse-vulnerability genes: genome scans for association, using 1004 subjects and 1494 single-nucleotide polymorphisms. *Am J Hum Genet* **69**: 1290–1300.

Vasiliskov AV *et al.* (1999) Fabrication of microarray of gel-immobilized compounds on a chip by copolymerization. *Biotechniques* **27**: 592.

Vo-Dinh T *et al.* (1999) DNA biochip using a phototransistor integrated circuit. *Anal Chem* **71**: 358–363.

Wallraff G *et al.* (1997) DNA sequencing on a chip. *Chemtech* (Feb) 22–32.

Wang DG *et al.* (1998) Large-scale identification, mapping, and genotyping of single-nucleotide polymorphisms in the human genome. *Science* **280**: 1077–1082.

Weiler J *et al.* (1997) Hybridisation based DNA screening on peptide nucleic acid (PNA) oligomer arrays. *Nucleic Acids Res* **25**: 2792–2799.

Yershov G *et al.* (1996) DNA analysis and diagnostics on oligonucleotide microchips. *Proc Natl Acad Sci USA* **93**: 4913–4918.

Zangenberg G *et al.* (1999) Multiplex PCR: optimization guidelines. In M Innis *et al.*, editors. *PCR Applications: Protocols for Functional Genomics*, pp. 73–94. San Diego, CA: Academic Press.

Protocol 5.1: Multiplexed allele-specific extension coupled with tag-arrays

MATERIALS

Reagents

Array preparation:

EtOH (absolute ethanol)

APTS (3-aminopropyl-triethoxysilane Sigma A-3648)

MeOH (methanol, analytical grade)

DMF (dimethylformamide, analytical grade)

Pyridine (Sigma P-3776)

PDITC (1,4-phenylene diisothiocyanate, Fluka 78480)

SuperClean slides (ArrayIT cat. no. SMC-25)

Array spotting:

Oligonucleotides with 3′ C6-aminomodifier

N,*N*-diisopropyl-ethylamine

Ammonium hydroxide (Sigma A-6899)

PCR:

Oligonucleotides with a 5′ T3 promoter sequence in the 'forward' primer (i.e. the primer that is in the same orientation as the detection primer):

TTCTAATACGACTCACTATAGGG

and a 5′ T7 promoter sequence in the 'reverse' primer:

TTCTAATACGACTCACTATAGGGAGA

AmpliTaq Gold buffer, Amplitaq Gold Enzyme and 25 mM $MgCl_2$ (Applied Biosystems cat.no. 4311806)

dNTP mix (dATP, dCTP, dGTP, dTTP)

Purified genomic DNA

Post-PCR target preparation:

Ampliscribe T7 High yield transcription kit (Epicentre cat. no. AS2607)

Primer extension:

MMLV reverse transcriptase, DTT and MMLV buffer (Epicentre cat. no. M6110K)

1 mM CY5-dUTP (Amersham-Pharmacia Biotech cat. no. PA55022)

D(+) trehalose (Sigma T-5251)

dNTPs (dA, dC, dG, dT)

Glycerol

Array hybridization:

Millipore MultiScreen HV plate (Millipore cat. no. MAHV N45 10)

Sephadex G-50 Superfine (Amersham-Pharmacia Biotech cat. no. 17-0041-01)

20× SSPE (3 M NaCl, 0.2 M NaH_2PO_4, 0.02 M EDTA)

50× Denhardts' (1% BSA, Ficoll and PVP)

Yeast tRNA (GibcoBRL cat. no. 15401-011)

Tween-20 (Sigma cat. no. P9416)

Equipment

Array spotting:

Contact (e.g. The Affymetrix® 427™ Arrayer, Affymetrix, Santa Clara, CA) or non-contact spotter (BioChip Arrayer, Perkin-Elmer Life Sciences Inc., Boston, MA)

PCR and primer extension: Thermocycler with 384-well block (such as GeneAmp PCR system 9700, Applied Biosystems)

Array hybridization:

Plate centrifuge (such as Sigma 4K15)

Hybridization oven/waterbath

Array read-out:

Confocal Laser scanner (such as ScanArray 5000XL, PerkinElmer Life Sciences Inc.)

Quantitation software (such as QuantArray 3.1, PerkinElmer Life Sciences Inc.)

METHODS

Primer design

Following selection of SNP-panel design PCR primers flanking the SNPs with low GC-content, minimal primer–dimer potential and short amplicon size (<130 bp) to facilitate multiplex PCR amplification use appropriate software for primer design (such as Primer 3.0, http://www-genome.wi.mit.edu/cgi-bin/primer/primer3.cgi) and apply the same criteria for all amplicons (e.g. same T_m, fragment length range). We design the PCR primers at 62°C melting temperature. To facilitate multiplex amplification (Wang *et al.*, 1998) and enable subsequent *in vitro* transcription (Pastinen *et al.*, 2000) all primer pairs have one primer with 5′ T3 promoter tail and another with 5′ T7 promoter (see Materials). The tag-oligonucleotides immobilized on generic arrays used in any given system should have minimal cross-hybridization potential and even melting temperatures, for design see (Gerry *et al.*, 1999 or Hirschhorn *et al.*, 2000). For each SNP two allele-specific detection primers are designed which differ by 3′ nucleotides complementary to 'variant' or 'normal' allele, each detection primer should have $T_m \approx 50$°C, and no internal hairpins. Each detection primer will be coupled with a different complementary tag-sequence based on the generic tag-array used. An example of a complete primer design for one SNP in the *APOB* gene (−837 G/A, antisense C/T change analysed) is shown below.

'Forward' PCR primer with a 5′ T3-promoter tail:

5′-ATTAACCCTCACTAAAGGGACAGGACACGTCATGTTCCTCATA-3′

'Reverse' PCR primer with a 5′ T7-promoter tail:

5′-TTCTAATACGACTCACTATAGGGAGAGGAAATGGGCAGTGCC-TAGAAG-3′

C-allele detection primer with 5′ cTag#1 sequence tail:

5′-TAGAGAGAGTCCACACACGCCATGCATCGTTTCCTTCC-3′

T-allele detection primer with 5′ cTag#2 sequence tail:

5′-AGCTCGACGTTCGGACACATCATGCATCGTTTCCTTCT-3′

Tag-sequence (Tag#1) to be immobilized on generic arrays:

5′-GCGTGTGTGGACTCTCTCTA-C6 Aminomodifier 3′

Tag-sequence (Tag#2) to be immobilized on generic arrays:

5′-ATGTGTCCGAACGTCGAGCT-C6 Aminomodifier 3′

Surface derivatization

The SuperClean slides do not require further cleaning, if standard glass slides are used a cleaning procedure prior to silylation is required (such as described by Lindroos *et al.*, 2001).

Silylation:

1. Immerse slides into EtOH/APTS/dH_2O solution (95:2:3 by volume).
2. Place slide container on a shaker (100 rpm) for 2 h.
3. Wash slides once with EtOH, MeOH and DMF (each wash 5 min on a shaker).

Activation:

1. Immerse silylated slides into 10 mmol PDITC solution in DMF supplemented with 1% (v/v) pyridine.
2. Incubate the slide container on a shaker (100 rpm) for 2 h.
3. Wash slides once with DMF and two times in MeOH.
4. Spin slides dry in centrifuge (1000 rpm, 2 min).
5. Store activated slides under nitrogen or in a desiccator until spotted (up to several weeks).

Spotting

Prepare 10 μM dilution of the amino-modified tags in 1% (v/v) *N*,*N*-diisopropyl-ethylamine in dH_2O and dispense the desired array pattern on activated slide surfaces. After spotting, rehydrate the spots in a humidified chamber for 24 h (the reaction between the aminated oligonucleotides and activated surfaces is slow), rinse slides in 1% ammonium hydroxide (v/v) and dH_2O. The spotted slides can be stored for a few weeks at –20°C.

Multiplex PCR

Prepare a primer mix with desired forward and reverse primers for each SNP panel to be screened. Carry out multiplex PCR in 384-well plates with 0.05 μM (final concentration) primer mix, 0.3 mM dNTPs, 2.0 mM $MgCl_2$, 10 ng of genomic DNA and 1 U of Amplitaq Gold enzyme in 5 μl of 1× Gold buffer. Apply thermocycling with initial activation at 95°C for 12 min, the following cycling parameters are: 96°C for 30 s, 66°C − 1°C per cycle for 4 min for six cycles; 96°C for 30 s, 60°C − 0.5°C per cycle for 2 min and 72°C for 2 min for 14 cycles; 96°C for 30 s, 54°C for 30 s, 72°C for 2 min for 20 cycles; 72°C for 6 min. We generally apply 24–48-plex PCRs, with successful amplification in 70–90% of the amplicons. Improved success in multiplex PCR can be achieved by initial verification that each amplicon amplifies individually in standard PCR.

Post-PCR target preparation

An *in vitro* transcription reaction is performed on multiplex PCR products to produce single-stranded RNA target for allele-specific primer extension and to further amplify the targets (Pastinen *et al.*, 2000). One microlitre of the PCR product is transcribed to RNA in a 2.5 μl *in vitro* transcription reaction using the Ampliscribe T7 high-yield transcription kit. Apply the reagent concentrations suggested by the manufacturer and incubate at 37°C for 2–4 h. Alternatively, the complete PCR product can be ethanol precipitated in 384-well plates and the *in vitro* transcription reaction can be carried out directly on the PCR plate.

Allele-specific primer extension reaction

The complete primer extension procedure is carried out on a thermocycler. Add 1 μl of the detection primer mix (0.5 μM with respect to each detection primer) to the *in vitro* transcription products (2.5 μl) and heat the RNA-detection primer mix to 85°C for 3 min, and cool to 50°C for primer annealing. Prepare 7 μl of primer extension mix per reaction containing: 0.9 M trehalose, 15% glycerol, 5 U MMLV enzyme, 10 mM dithiothreitol (DTT), 1× MMLV buffer, 75 μM dA/dC/dG mix, 5 μM dT and 2.5 μM CY5-dU. Add 7 μl of the primer extension mix to the RNA-detecting primer mix at 50°C and suspend to mix, seal plate (using silicon covers) and increase the temperature to 55°C for 45 min. The primer extension products are transferred on ice following the incubation. We routinely multiplex 48 SNPs (96 detection primers) in the extension reaction (i.e. the same level of multiplexing as in PCR), but have successfully multiplexed up to 80 SNPs (160 detection primers). Furthermore, we use four different dyes in parallel for different samples in the primer extension phase (Cy5, Cy3, BodipyFL and Texas Red) to increase throughput at subsequent steps of the procedure.

Purification and tag-array hybridization

Prepare the Multiscreen-Sephadex purification plates as suggested by the manufacturer (Millipore Tech Note TN 050, Millipore, Bedford, MA). If the primer extension reaction was carried out using four dyes, one reaction from each dye can be combined prior to loading to purification plates (i.e. each purified sample will then be composed of four different samples each labelled with a different dye), if the volume to be loaded on purification plates is <20 μl the volume should be increased to 20 μl with dH_2O to ensure recovery of the products. Prior to hybridization of the purified products add 20× SSPE, 50× Denhardts' and yeast tRNA to final concentrations of 4×, 2× and 0.1 mg/ml, respectively. Concentrated solutions of the hybridization buffer are used to prevent excessive dilution of the sample prior to hybridization. Denature samples at 95°C for 3 min and transfer on ice for loading on the arrays. If large sample sizes are processed a practical solution for hybridization is the use of custom-made silicon-rubber grids to enable application of up to 80 hybridization solutions per slide ('array-of-arrays', Pastinen *et al.*, 2000). Alternatively, any commercially available hybridization chamber designed for expression analysis can be used. Good results are obtained using 8–20 μl of hybridization solution per array (depending on the volume of the chamber used) and hybridization at 48°C (T_m of the tag-sequences 52–60°C) for 1 h followed by 30 min washes in 6× SSPE, 0.01% Tween-20 and 3× SSPE, 0.01% Tween-20 at 48°C. After the second SSPE wash the slide is rinsed with ice-cold dH_2O and centrifuged dry (1000 rpm for ≈3 min).

Array scanning, signal quantitation and result interpretation

We routinely apply four-colour hybridization and subsequent scanning using four channels, currently the only confocal scanners allowing this are the ScanArray 5000 (four excitation wavelengths) and ScanArray 5000XL instruments (five excitation wavelengths). Our spots are 90–120 μm in diameter and scanning is carried out using either 5 μm or 10 μm resolution. A scanning result from our tag-arrays is shown in *Figure 5.5*. The 16-bit TIFF-images are quantified using the QuantArray 3.0 software, applying the 'fixed-circle' quantitation method and local background subtraction. The genotype clusters can be visualized by plotting log 2 transformed signal intensity values from the two tags corresponding to the alleles on *x*- and *y*-axes. A result of genotyping the APOB-837 SNP in a 48-plex reaction is shown in *Figure 5.6*. Automated calling of the alleles could be achieved by applying clustering algorithms identifying the genotype clusters, and we are currently working (in collaboration with Dr AD Long, UCI, CA) to develop software to automatically normalize (based on the multi-channel data) signal intensities and call genotypes. The procedure described here yields 70–80% genotyping success from completely multiplexed scoring reaction for 48 SNPs in parallel, and miscall rates with appropriate quality control can be <1% in large-scale applications.

Microplate-Based mRNA Analysis

6

Taku Murakami and Masato Mitsuhashi

6.1 Introduction

The Human Genome Project has uncovered ≈30 000 genes from human chromosomes (Lander *et al.*, 2001; Venter *et al.*, 2001). Each gene will be transcribed to mRNA, but the species of expressed mRNA and their expression levels vary among different types of cells and tissues, and also differ among cell cycles and ambient environments. Even within the same cell type, mRNA expression is regulated exquisitely in response to various environmental changes, cell-to-cell communications and extrinsic/intrinsic stimulations. Thus, mRNA is a principal material for gene expression analysis, cDNA cloning, gene discovery project, etc.

Gene expression is the sum of transcription, splicing, mRNA transport from the nucleus to the cytoplasm, protein translation, mRNA degradation and the post-translational modification of proteins. Because proteins are major players in biological activities, the qualitative and quantitative analyses of the expressed proteins are prime targets for researchers in bioscience and pharmaceutical drug discovery projects. However, because of limitations in protein analysis technologies, mRNA will supplement some protein analysis. Unlike protein analysis, in which specific antibodies and/or detailed structural/functional information are required, mRNA can be analysed whenever its sequence information becomes available in GenBank or other public and private gene sequence databases. Because Northern blot analysis (Sambrook *et al.*, 1989) and PCR (Saiki *et al.*, 1985) are able to identify and quantitate the levels of specific mRNA even from a pool of different genes, the specific mRNA of interest can be analysed without purification from other genes if gene-specific primers and probes are available. Moreover, because stability is different for each protein, no universal preservation method is available for proteins, whereas RNA can be preserved by inactivating ribonuclease activity. In order to analyse protein expression in biological specimens, 2D gel electrophoresis is frequently used (O'Farrell, 1975). However, this requires large quantities of proteins for analysis, and unlike DNA and RNA, no amplification technology is available for proteins.

DNA microarray chip technology is a very powerful tool for gene expression analysis (Schena *et al.*, 1995). It allows the simultaneous screening of many genes from a single specimen. However, it requires large amounts of starting RNA or mRNA. The costs of sample preparation and of the chip itself are prohibitory factors for some laboratories. The quality control of thousands of immobilized probes is another issue for chip manufacturers.

Microarrays & Microplates: Applications in Biomedical Sciences, Shu Ye and Ian N.M. Day

Moreover fluorescent signals are less sensitive than gene amplification technologies. Therefore, microarray technology may be suitable for the first screening of genes of interest, whereas microplate-based high-throughput assays will be suitable for the second screening and the precise analysis of mRNA.

6.2 Microplate-based RNA purification

Purification of clean and intact RNA is the essential first step in gene expression analysis. The first procedure of RNA purification releases RNA from cell bodies, and is usually performed by adding lysis reagents to samples and mixing vigorously. If the samples are solid materials such as tissues or organs, the mixing procedure is quite labour-intensive, and is not suitable for microplate-based high-throughput assays. However, if the target samples are cultured cells grown in microplates, one should consider a microplate-based RNA purification system, because the lysis procedure can be conducted by adding lysis reagents to the microplate wells. The function of lysis reagents is to deactivate endogenous RNase to prevent RNA degradation, and to break down the lipid bilayers of cell membranes to release RNA from cells. Different lysis reagents with different optimizations for each sample are commercially available, for example, for cultured cells, blood, plants and animal tissues.

If the sample to be analysed is made up of adherent cells, a microplate-based assay is an ideal platform because culture medium can be removed by aspiration, followed by the addition of lysis reagent to each well. However, the lysis of floating cells is more complicated than that of adherent cells, because it requires additional steps to separate the floating cells from the culture medium. One of the simplest methods is to add concentrated lysis reagent directly to the culture medium. However, some medium contains more RNases than the lysis reagent can inactivate. It may contain unknown factors to inhibit RNA purification. If the volume of the culture medium is 200 μl in an ordinary 96-well microplate, no more lysis reagent can be added to the wells because of physical limitations.

There are two methods of separating floating cells from culture medium. The first is to precipitate the culture cells by centrifugation. However, this requires a special rotor to hold the microplates. Moreover, it is not easy to completely remove the culture medium without disturbing the precipitated cultured cells. In order to lyse the precipitated cells, pipetting or vortexing is unavoidable. However, because the release of RNA from the mass of precipitated cells is dependent of the strength of the mechanical force of pipetting or vortexing, equal mixing techniques are required to minimize sample-to-sample variation. The second method is to use filter plates. Using filter plates, floating cells are trapped on the filter membranes and washed with appropriate solution to remove materials undesirable for the downstream procedures. Filtration can be carried out by either centrifugation or vacuum aspiration.

The next step is to isolate RNA from other cellular components, such as genomic and mitochondrial DNA, lipid membrane, proteins, etc. RNA can be purified as total RNA or poly(A)$^+$ RNA. Total RNA is widely used as a material for gene expression analysis or other RNA experiments, because purification of total RNA is much easier than poly(A)$^+$ RNA. However, it is not always true, and the recent procedure of poly(A)$^+$ RNA purification is much simpler and easier than total RNA (Miura *et al.*, 1996). Total RNA is purified with phenol–chloroform extraction followed by ultracentrifugation across a caesium chloride gradient (Glisin *et al.*, 1974; Palmiter, 1974). Ultracentrifugation can be replaced by DNase treatment to remove contaminated DNA. RNA is also purified by acid-guanidine/phenol/chloroform (AGPC) extraction (Aso *et al.*, 1998). However, these methods are not acceptable for microplate-based high-throughput assays. Alternatively, RNA is adsorbed using silica particles, washed with appropriate buffer, and eluted with hypotonic solution. However, mRNA accounts for only 1–5% of total RNA, and >95% of total are rRNA and tRNA. This rRNA and tRNA may induce high background in some applications. Normalization of samples is troublesome if total RNA is used, because equal amounts of total RNA do not guarantee that the amounts of mRNA are equal. The alternative is the use of control genes such as β-actin (Ponte *et al.*, 1983) and glyceraldehyde-3-phosphate dehydrogenase (GAPDH; Tso *et al.*, 1985). However, expression of these genes is also known to vary substantially under certain conditions. If cDNA is transcribed from total RNA, reverse transcriptase will use not only mRNA, but also rRNA and tRNA, as templates. As a result, cDNA will contain many false transcripts from rRNA or tRNA. In differential display technology (Liang & Pardee, 1992), false amplification due to rRNA and tRNA will give a high background and decrease sensitivity on gel analysis.

Highly purified mRNA is recommended in many applications, because it improves the sensitivity and specificity of the assay. Although mRNA can be purified with oligo(dT)-immobilized solid support like as poly(A)$^+$ RNA, this method is not applicable to bacteria and certain types of viruses, because of the lack of poly(A)$^+$ tail. If endogenous RNases digest mRNA, 3′ fragments of mRNA can be purified by oligo(dT)-immobilized solid support, however, 5′ fragments of mRNA cannot. Thus, if poly(A)$^+$ RNA is used for quantitative gene expression analysis, probes or primers should be located at the 3′-end.

Recently, various types of RNA purification products have become available. These products allow simpler, faster and safer RNA purification protocols than organic solvent-based extraction methods. Three product categories allow high throughput assay and automation (*Table 6.1*, *Figure 6.1*). The first is to use silica- or glass-based membrane filters. This method utilizes the feature that RNA is adsorbed selectively to either silica or the glass surface of the membrane filters. Using filter plates with the same dimension as the microplates, 96 samples can be processed simultaneously, and RNA is eluted by

Table 6.1 Automation-compatible RNA purification

Format	Type of RNA	Amount of purified RNA	Sample-to-sample variance	Special requirements	Major suppliers
Membrane filter plate	Total RNA	Large	High	Microplate centrifuge, vacuum aspirator, etc.	RNAqueous™-96 (Ambion), RNeasy® 96 Kits (QIAGEN), etc.
Magnetic beads	mRNA	Variable	High	Magnetic separator	Dynabeads® (Dynal Biotech), MPG® (CPG Inc.), etc.
Oligo(dT) microplate	mRNA	Middle	Low	–	GenePlate® (RNAture)

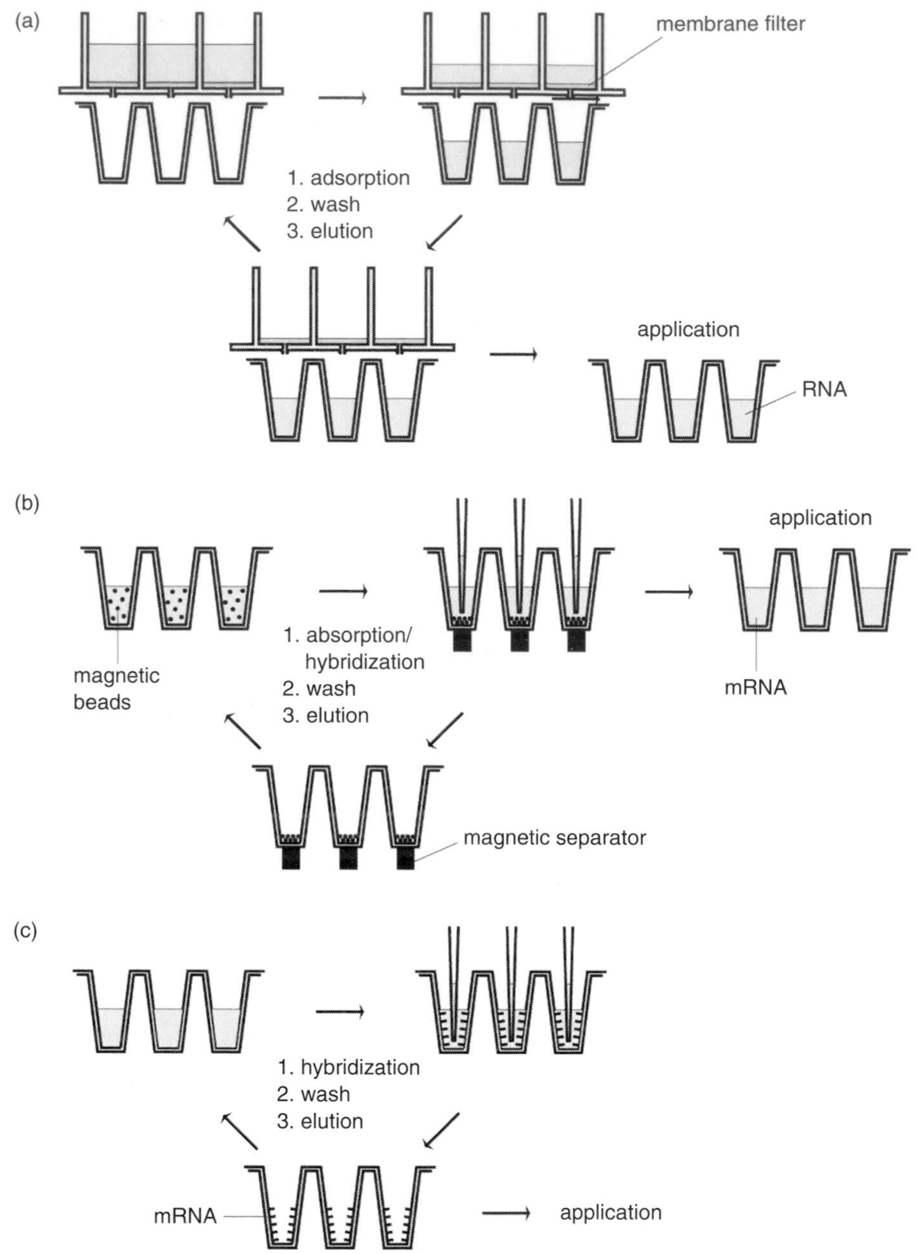

Figure 6.1

Comparison of RNA purification protocols. (a) Membrane filter plate. This system consists of two microplates. Cell lysates and buffers for wash or elution are applied to the top plate, the bottom of which contains silica- or glass-based membrane filters. Each buffer is forced through the membrane to the bottom plate by centrifugation or vacuum aspiration at each stage of hybridization, wash and elution. RNA is adsorbed to the membrane when cell lysates pass through the membrane. After undesired materials have been removed by the passage of wash buffer, adsorbed RNA is eluted with elution buffer into a newly prepared plate. (b) Magnetic beads. This system consists of a microplate, magnetic beads and a magnetic separator. Because RNA is adsorbed or hybridized to the surface of the magnetic beads, cell lysates and buffers can be removed by pipetting following collection of magnetic beads. RNA is eluted into the elution buffer, which can be transferred to a fresh microplate. (c) Oligo(dT)-immobilized microplate. This system consists of an oligo(dT)-immobilized microplate only. RNA is adsorbed to the well surface, thus cell lysates and buffers can be removed by pipetting only. RNA is eluted into the elution buffer or can be used directly in the downstream assays on the plate.

adding hypotonic buffer (*Figure 6.1a*). However, this technology is only applicable to total RNA, and carries all the disadvantages of total RNA as described above (*Table 6.1*). Moreover, it requires centrifugation or vacuum aspiration, and well-to-well variation is larger than with other methods (*Table 6.1*).

The second method is a magnetic beads-based technology (*Table 6.1, Figure 6.1b*). Silica surface or immobilized oligo(dT) on the magnetic beads allows the capture of total RNA or poly(A)$^+$ RNA, respectively. Magnetic separation of the beads is applicable to microplate formats such as 96-, 384- and 1536-well plates. Captured RNA or poly(A)$^+$ RNA is released from the magnetic beads by hypotonic elution buffer. However, it requires dedicated instruments for automation (*Table 6.1*). If the magnet is too strong, the beads form a clump and are difficult to disperse. If this happens, wash efficiency is reduced substantially. Furthermore, some portion of the magnetic beads may be removed during magnetic separation, inducing sample-to-sample variation (*Table 6.1*).

The third method uses oligo(dT)-immobilized microplates (*Table 6.1, Figure 6.1c*) (Mitsuhashi *et al.*, 1992; Hamaguchi *et al.*, 1998). This technology provides the simplest mRNA purification protocol. Oligo(dT) is immobilized on the well surface of microplates by physical adsorption (Yamazaki *et al.*, 1993), covalent bonding (Taniguchi *et al.*, 1994), biotin–streptavidin or antigen–antibody conjugation (Jalava *et al.*, 1993), or synthesis of oligonucleotide on the surface of microplates (Chee *et al.*, 1996). Researchers simply add cell lysates to the oligo(dT)-immobilized well, incubate at room temperature for hybridization, wash each well with appropriate buffer to remove non-hybridized materials, and elute the purified mRNA from each well using hypotonic elution buffer (*Figure 6.1c*). This process does not require any special or expensive devices such as a centrifuge or vacuum manifold, and can be performed by manual operation with a multi-channel pipette or commonly available auto-dispensing instruments (*Table 6.1*). Thus, this technology can be automated with ordinary liquid-handling instruments (*Figure 6.2*) (Mitsuhashi, 2000). Moreover, because the surface area of each well is quite similar, well-to-well variation can be minimized. Although the surface area is much smaller than that of filter plates or magnetic beads, it may be compensated for by sensitive downstream assays such as PCR.

6.3 Microplate-based PCR

Several fundamental techniques such as Northern blot analysis (Sambrook *et al.*, 1989), ribonuclease protection assays (RPAs; Zinn *et al.*, 1983; Melton *et al.*, 1984) and reverse transcription polymerase chain reaction (RT-PCR; Kawasaki & Wang, 1989; Wang *et al.*, 1989) have been developed and used to measure mRNA expression levels in samples. Among them, RT-PCR exhibits the highest sensitivity and is capable of detecting a single copy of the transcript. At the first step of RT-PCR, mRNA is reverse transcribed to cDNA using reverse transcriptase. The synthesized cDNA is then used as a template for PCR. Recently, this RT-PCR technology has been performed in microplates because the 96- and 384-well formats of thermal cyclers and the corresponding heat-stable microplates have become commercially available. Because RNA or mRNA is purified in a microplate as described above, whole processes of RT-PCR are now conducted in microplates, making microplate-based high-throughput RT-PCR realistic.

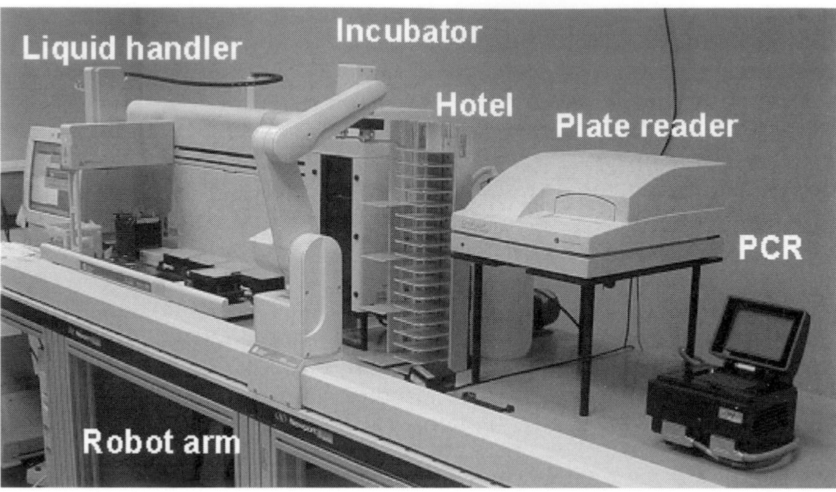

Figure 6.2

Automation platform of mRNA analysis. mRNA is purified in a liquid handler, cDNA is synthesized in an incubator, followed by PCR in a robot-friendly thermal cycler. Post-PCR analysis is prepared in a liquid handler, and the fluorescent signal is detected in a microplate reader. A robot arm moves the microplate from one station to another. Using a stack of plates in a hotel, multiple plates are processed sequentially.

When RNA or mRNA is eluted from the solid surface at the final step of RNA purification, it is dissolved in hypotonic buffer or water. Thus, one can simply mix cDNA synthesis reagents with RNA/mRNA to start RT-PCR. It is not necessary to adjust pH and salt concentrations. However, if RNA/mRNA is diluted below the detection limit, it should be concentrated by lyophilization, Speed Vac or ethanol precipitation. These concentration methods are time-consuming and require special microplate-holders for centrifugation.

The buffer compositions for RT and PCR are slightly different. Thus, some residual components of the RT reaction mixture change the composition of the PCR mixture and some by-products of cDNA synthesis may also inhibit PCR. Therefore, the synthesized cDNA should be diluted substantially until these adverse effects are eliminated. If large amounts of cDNA are used for a highly sensitive assay, cDNA should be precipitated using ethanol, although ethanol precipitation is not suitable for microplate assays.

One-step RT-PCR, in which RT and PCR take place sequentially in a single well, may overcome some of these issues. One-step RT-PCR does not require dilution of the cDNA, and whole cDNA is used for PCR. Therefore, one-step RT-PCR is suitable for microplate-based high-throughput assays with minimum cross-contamination between sample transfer. However, because the optimum buffer conditions, in which enzymes have their maximum activities and primers hybridize to the target sequences stringently, are different between RT and PCR, the sensitivity and specificity of the entire assay may decrease in comparison with regular two-step RT-PCR. More importantly, primers bind at low stringent conditions for longer during RT, and the resultant partial primer–primer hybrids will become double-stranded DNA by the activity of DNA polymerase. Once primer dimers are formed, these are no longer available to PCR, and reduce the assay sensitivity (Hamaguchi

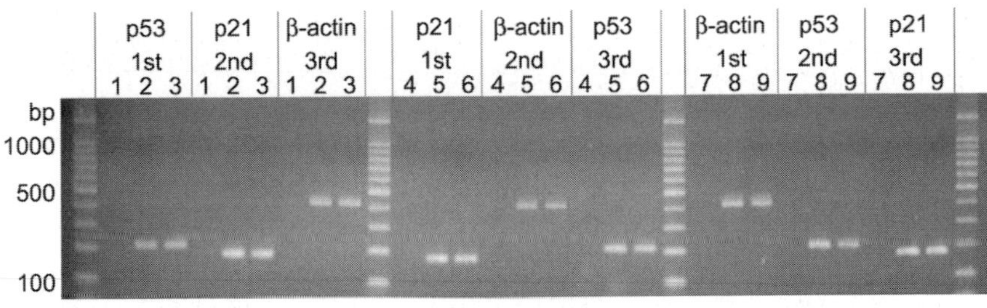

Figure 6.3

Sequential PCR in an oligo(dT)-immobilized microplate. HEK293 cells (ATCC, Manassas, VA) were mixed with lysis buffer (RNAture, Irvine, CA) and transferred to an oligo(dT)-immobilized microplate (GenePlate, RNAture) for hybridization at room temperature for 1 h. After non-hybridized materials had been removed by aspiration, cDNA was synthesized in the same GenePlate by adding MMLV reverse transcriptase and appropriate buffer and mononucleotides and incubating at 37°C for 1 h. Each well was washed with Wash Buffer (RNAture), and p53, p21 and β-actin were amplified by PCR in the same well sequentially with three different orders (1st, 2nd and 3rd). Experiments 2, 3, 5, 6, 8 and 9 were duplicate experiments and 1, 4 and 7 were negative controls without cell lysates.

et al., 1998). Furthermore, because synthesized cDNA cannot be aliquoted for multiple PCR reactions, the one-step RT-PCR method is not economical for multiple gene analysis from the same RNA samples.

Oligo(dT)-immobilized microplate technology (Mitsuhashi *et al.*, 1992; Hamaguchi *et al.*, 1998) solves some of these problems. After mRNA is captured on the microplate surface, cDNA can be synthesized in the same wells by simply adding RT reaction mixture, using immobilized oligo(dT) as an RT primer. After cDNA synthesis, buffer can be changed by aspirating cDNA buffer and dispensing PCR reagents into the wells, because the synthesized cDNA is immobilized to the microplates. Because entire mRNA is used for RT-PCR under optimal conditions for both RT and PCR, this assay exhibits maximal sensitivity (Hamaguchi *et al.*, 1998). Moreover, because PCR products are produced in solution, and template cDNA is kept immobilized during PCR, PCR can be repeated many times sequentially in the same wells using different PCR primers (*Figure 6.3*) (Hamaguchi *et al.*, 1998). If the control gene is amplified from the wells in which the target gene was amplified previously, the results are more reliable than the assays with two different reactions. Because cDNA is much more stable than mRNA, the immobilized cDNA will become cDNA bank for storage and archives.

Oligo(dT)-immobilized microplates can be used for multiple gene analysis, by synthesizing cDNA with random hexamers as primers. The resultant cDNA exists in solution, and is used simultaneously for multiple PCR. If 96 cDNA samples are prepared in 96-well microplates, and each cDNA is dispensed into 4 different wells of 384-well PCR plates, 4 different genes can be analysed simultaneously from 96 samples.

If the target materials do not contain poly(A)$^+$ tails, such as bacterial RNA and viral genomic RNA, gene-specific oligonucleotide can be immobilized to the microplates. Once target RNA is captured on the microplate, various assays become available as shown in the oligo(dT)-immobilized microplate.

PCR products are usually separated by agarose gel electrophoresis and stained with DNA-specific fluorescent dyes. Although this assay is simple for small numbers of samples, it becomes unrealistic when microplates produce 96 or 384 samples simultaneously. Thus, for high-throughput assays, PCR products are analysed in microplates where sequence-specific oligonucleotides are immobilized. Double-stranded PCR products are heat denatured, and applied to the microplates for hybridization, followed by the quantitation of hybridized PCR products with a variety of detection chemistries [reverse dot blot or post-PCR enzyme-linked immunosorbent assay (ELISA)]. However, a major obstacle is the manipulation of amplified PCR products, which frequently induces contamination into the environment. Furthermore, if PCR is stopped prematurely, samples with a low copy number of the target gene are not detectable, whereas extended PCR cycles can detect such samples, but samples containing abundant genes are difficult to quantitate, because PCR products reach a plateau. Thus, optimal PCR cycles should be identified for each assay.

In order to overcome the problems of agarose gel electrophoresis and post-PCR assays, real-time PCR has been developed and has become the mainstay of recent gene quantitation assays. It quantitates the amounts of PCR products continuously during PCR cycles, and accepts 96- or 384-well PCR plates. Various detection chemistries are available for real-time PCR, such as Taqman (Livak *et al.*, 1995), molecular beacon (Tyagi & Kramer, 1996) and SYBR green methods (Wittwer *et al.*, 1997a, b), and many companies are producing the suitable instruments. Because PCR reaction mixtures are enclosed during PCR, and amplified PCR products are never exposed to open environment in real-time PCR, it prevents contamination of amplified PCR products. When serial dilutions of the standard gene are included in the same microplates, the amounts of mRNA in tested samples can be quantitated automatically, even if the amounts of the target genes vary widely among samples.

Although microplates are ideal vessels for high-throughput automation, current PCR microplates are not ideal for automation, because of the non-rigidity, shrinkage and warp after PCR, and some auto-fluorescence. Polystyrene microplates are standard materials for automation, however, polystyrene is too heat sensitive to apply to PCR. Thus, heat-stable polypropylene is used for PCR, but it carries the problems described above. The most critical problem is a warping or bending of the microplates after PCR. If microplates warp, it is difficult to remove the PCR products from each well, because the *z*-axis location of the wells is not consistent. In real-time PCR applications, the warp induces changes in the light path. The warp becomes more problematic when 384-well microplates are used. There are three approaches to address warping problems. Some instrument companies produce a special platform, which mechanically holds microplates flat during PCR. Some plastic manufacturers use dual-injection moulding techniques to produce polypropylene wells with rigid frame. Our research group recently identified unique plastic compounds, which exhibit heat stability for PCR, rigidity similar to polystyrene, good heat conductivity similar to polypropylene, very low water absorption and auto-fluorescence, with a precision-injection moulding capability to make thin walled wells (*Table 6.2*). When we made 96- and 384-well microplates using these compounds, we found that the PCR performance was identical to that of polypropylene. Moreover, these microplates did not warp after PCR.

Table 6.2 Automation-compatible PCR microplate

	Ideal plate	Conventional plate
Rigidity	Rigid	Soft
Shrinkage	No	Slight
Robot handling	Easy	Difficult
Manufacturing flexibility	96, 384, 1536 well	96, 384 well
Wall thickness	<0.3 mm	0.3 mm
Heat resistance	$\gg 110°C$	$\geqslant 110°C$
Heat conductivity	$>2.8 \times 10^{-4}$ cal cm^{-1} s K	2.8×10^{-4} cal cm^{-1} s K
Heat deformation	No	Bend, warp, etc.
Transparency	Transparent	Semi-transparent
Fluorescent background	Ultra-low (ultrasensitive detection)	Low

6.4 Microplate-based non-RT PCR assays

PCR was the first technology for gene amplification, however, other gene amplification technologies are available, such as nucleic acid sequence-based amplification (NASBA; Kievits *et al.*, 1991), strand displacement amplification (SDA; Walker *et al.*, 1992a, b), etc. Although these technologies are more complicated than PCR, these are isothermal reactions, and easily applicable to microplate-based assays. Once RNA or mRNA is purified from microplates, these are immediately applicable to microplate-based gene amplification to detect specific genes. However, if the detection of amplified products is agarose gel electrophoresis or post-amplification assays, it carries the same problems described for PCR. Thus, real-time PCR is now implemented in these gene amplification technologies (Walker *et al.*, 1996; Leone *et al.*, 1998).

Attempts were made to detect mRNA using signal amplification, without amplifying genes. Once specific oligonucleotides are immobilized to microplates, the corresponding mRNA is captured from crude cell lysates. The resultant captured mRNA is hybridized with labelled probes (sandwich hybridization), or used to synthesize cDNA in the presence of labelled nucleotides (Tominaga *et al.*, 1996). The incorporated labelled compounds can be detected using colorimetric, fluorescent or chemiluminescent assays. In order to amplify the signals of labelled compounds, branched DNA (bDNA) amplification (Urdea *et al.*, 1991), cycling probe technology (CPT; Duck *et al.*, 1990), or Invader assay (Lyamichev *et al.*, 1999) can be used. However, because signal amplification is a linear reaction except for the bDNA assay, the sensitivity is much less than that of gene amplification technologies.

Oligo(dT)-immobilized microplate technology can be used to amplify RNA, if starting materials are limited (Mitsuhashi *et al.*, 1992). For example, RNA polymerase promoter sequences can be attached at the 5′ portion of immobilized oligo(dT) on the microplate. After mRNA is hybridized with oligo(dT), double-stranded cDNA is synthesized on the microplate, where RNA polymerase promoter region becomes double stranded. When RNA polymerase, rNTP, and appropriate buffer are added to the microplate wells, cRNA is transcribed *in vitro*, where a single copy of template ds-cDNA produces multiple copies of cRNA (Mitsuhashi *et al.*, 1992). If labelled rNTP is used, the resultant cRNA can be applicable to DNA microarray chips.

Recently, a DNA microarray chip has used microplates, by immobilizing multiple probes at the bottom of each well (Martel, 2001). Although the detection sensitivity of each spot is less than RT-PCR, it may be the future platform for multiple gene detection from multiple samples.

Starting materials for gene expression analysis or mRNA quantitation are always total RNA or poly(A)$^+$ RNA as described above. The result is thus the sum of transcription and degradation of mRNA. Increases in specific mRNA signals following stimulation indicate that new mRNA has been transcribed, the degradation rate has decreased, or a combination of the two. If degradation is enhanced, increases in transcription are difficult to identify. Additionally, when a large quantity of mRNA is present in the cells, small increases in transcription are overlooked. The nuclear run-on assay is used to measure the level of transcription of specific genes (Okamoto *et al.*, 1994). However, it is not suitable for microplate-based high-throughput assays. Thus, we recently developed a microplate-based assay for transcription analysis (Matsuda *et al.*, 2002). In this method, cells are trapped on the glass-fibre membranes of 96-well filter plates and subsequently exposed to non-ionic detergent to achieve cell membrane permeation. The cytosolic components, which contain pre-existing mRNA, are removed by washing extensively with appropriate buffer, while nuclei remain in the filter plates (Matsuda *et al.*, 2002). Lysis buffer is then used to release nuclear mRNA, which is collected on oligo(dT)-immobilized PCR plates for the capture of poly(A)$^+$ RNA, upon which RT-PCR is performed as described above. By removing pre-existing mRNA from the cytosol, and capturing newly synthesized poly(A)$^+$ RNA from the nucleus, the assay becomes very sensitive to some gene expression that cannot be detected using conventional whole-cell methods (Matsuda *et al.*, 2002).

6.5 Application of mRNA analysis to drug discovery project

In order to discover seed compounds for candidate drugs, large numbers of compounds need to be screened. If the function of the candidate drugs is known or speculated, this function is used for screening. The cell death or cell growth assay is a typical example. However, the functional assay is limited to specific area, where drug action can be mimicked *in vitro*. If drug candidates react with receptors, binding assay with native and/or recombinant receptors is widely used. However, developing a screening system for each receptor requires a lengthy, labour-intensive process of cloning, expression, purification of soluble receptors, and labelling of lead compounds. Immunoassay or ELISA (Engvall *et al.*, 1971) is also used in drug screening. However, the assay depends on the availability of specific antibodies. In any case, the microplate-based assay with 96-, 384-, 1536-well plates or much higher number of wells is a desirable format for analysing many samples.

Recently, gene expression analysis is also used as a screening tool for drug screening. DNA microarray chips are used to characterize the expression profile of multiple genes before and after drug treatments. However, DNA microarray chips still have many problems: for example, these chips are expensive, they require large amounts of mRNA/cDNA, and exhibit poor reproducibility. Once promoter sequences of the particular target gene have been identified, luciferase gene or some marker genes can be ligated at the end of the promoter sequences (Alam & Cook, 1990). If these recombinant genes are functionally expressed in the cell system, the action of drugs can easily be identified by measuring the signal of marker gene. However, this technique is only applicable to the genes of which promoter sequences are well characterized.

If mRNA is purified from cells, cDNA is synthesized, and the levels of specific gene are quantitated by PCR, such a system will become universal for gene expression-based drug screening. Because of the sensitivity of gene amplification technologies, expression of a candidate gene can be quantitated from a much smaller number of cells than is required by DNA microarray chip analysis. Furthermore, once constructed, the same system can be applied to many drug targets, simply by changing the cell types and amplification primers. Primers can be designed from GenBank or cloned gene sequences using appropriate software (Mitsuhashi *et al.*, 1994; Hyndman *et al.*, 1996). However, because the traditional mRNA preparation method does not have a microplate format, this type of assay was not used in drug screening until recently. Fortunately, a microplate-based purification system for total RNA or mRNA is commercially available and a corresponding automation system has been developed, therefore gene expression-based drug screening has become a reality. Furthermore, an identical system can be used not only for drug screening, but also for toxicology tests if toxicity-related genes are characterized.

Microplate-based gene expression analysis is also useful for gene discovery projects or functional genomics. The Human Genome Project had predicted ≈30 000 genes in human chromosomes (Lander *et al.*, 2001; Venter *et al.*, 2001). However, the function of many genes remains unknown. In order to analyse the function of genes, one can inhibit individual gene expression by administering gene-specific compounds, such as antisense oligonucleotides (Toulme & Helene, 1988) and ribozymes (Perriman & Gerlach, 1990) for mRNA cleavage, or dumbbell oligonucleotides (Clusel *et al.*, 1993) for transcription inhibitor. mRNA analysis is used as the first step to confirm whether the target gene expression is really inhibited.

6.6 Application of mRNA analysis to clinical diagnostics

Within the same individual, DNA sequences are essentially identical among cells, tissues and organs. DNA is also much more stable than mRNA. Thus, molecular diagnostics first became available for DNA to identify mutation or SNP for genetic diseases, cancer and drug responsiveness. DNA diagnostics is also used for infectious diseases to identify and quantitate foreign DNA. However, in contrast to DNA, mRNA varies among cells, tissues and organs, and even in the same tissues, mRNA differs between normal and disease status, and changes rapidly in response to treatment. Thus, mRNA diagnostics provides functional information of the particular disease site at the time of the test. mRNA diagnostics is a phenotyping test using genetic information and biotechnologies.

One of the major obstacles to mRNA diagnostics is the presence of multiple cellular components in biomedical specimens. Because mRNA expression differs among cells, the existence of vascular cells (endothelial cells, smooth muscle cells), blood cells (leukocytes), nerve/synapse and mesenchymal cells (fibroblasts, astrocytes) in the biopsy or surgical specimen makes the results difficult to interpret. Whole blood contains granulocytes and lymphocytes. Interestingly, platelets also contain mRNA (Newman *et al.*, 1988). Even in the mononuclear cell fraction, there are many subsets of lymphocytes.

However, because gene amplification technologies can identify and quantitate the levels of specific mRNA, mRNA diagnostics can be realized if the specific gene(s) is expressed only in the target cells. One example is the detection of leukaemia cells in blood (Kawasaki *et al.*, 1988). Some leukaemia cells exhibit translocation, and the resultant chimaera mRNA derived from the translocated DNA lesion exists only in leukaemia cells. Using gene amplification technology, a single leukaemia cell can be detected from a pool of normal cells by amplifying a leukaemia-specific gene. This type of diagnostics is widely available in academic institutions as home-brew diagnostics.

Another example is micrometastatic cancers. Colon, prostate and thyroid cancers, and melanoma express unique genes, such as carcinoembrionic antigen (CEA), prostate-specific antigen (PSA), thyroglobin and tyrosinase, respectively (Pelkey *et al.*, 1996). Solid cancers also express cytokeratin 19 (Ferrero *et al.*, 1990). Because these genes are not expressed in blood cells, the existence of these genes indicates the presence of cancer cells in the blood (so-called micrometastasis). Unlike X-ray, CT-scan, NMR, endoscopy, ultrasound, scintillation scan, etc., the blood test is not invasive, and is suitable for mass screening. Thus, we believe that the quantitation of cancer-specific mRNA in blood may be applicable to future cancer screening. Furthermore, if the amounts of cancer-specific mRNA correlated with the amounts of circulating cancer cells, this test will be useful for monitoring treatment and follow-up. For example, when surgery is successful or chemotherapy effective, cancer-derived mRNA will disappear from the blood. Relapse may be identified at an early stage if blood tests are performed periodically.

Lung cancer is also a target for mRNA diagnostics. Because lung is a highly vascularized tissue, lung cancer exhibits the highest rate of blood-borne metastasis of any cancers. Also blood-borne metastasis often happens even from a small mass which is difficult to visualize by X-ray analysis. Unfortunately, few specific gene markers are available for lung cancer. However, we have recently identified four marker genes for lung cancer, using extensive differential display analysis followed by RT-PCR screening (Matsunaga *et al.*, 2002). By combining these four markers and cytokeratin 19, the diagnostic value is increased dramatically. In fact, more than three-quarters of advanced-stage lung cancer patients expressed at least one of these genes in blood, whereas just one-twentieth healthy volunteers expressed one gene (Matsunaga *et al.*, 2002). Because lung cancer does not have a suitable screening procedure for early detection, and exhibits the highest mortality rate of any cancer in industrial countries, we hope that this type of mRNA diagnostics will soon be applicable in clinical practice. Because health screening tests should process many patients' specimens simultaneously, a microplate-based assay platform is highly desirable.

If patients are administered strong medication, drug sensitivity or responsiveness may be quantitated by measuring the expression of drug responsive genes. Because gene expression is a functional marker for drug responsiveness, it may be more meaningful than measuring drug concentrations in blood. It is also useful for individualized medication, because each patient may react differently. Furthermore, if blood is incubated with appropriate concentrations of candidate drugs *in vitro*, drug sensitivity may be identified by measuring the changes in expression of drug responsive genes in individual patients. This is a kind of futuristic 'tailored medicine' to prevent adverse effects. Genotyping tests to identify SNP are a promising method for tailored medicine, but mRNA-based phenotying tests may be complementary to the genotyping tests.

If the quantitation of specific mRNA expression is used for clinical diagnostics, the most critical factor is reproducibility: sample-to-sample, day-to-day or institution-to-institution variation. Because mRNA diagnostics requires multiple time-consuming steps, such as mRNA purification, cDNA synthesis, PCR and quantitation, any variation at each step will be substantial by the end of the process. In particular, gene amplification technology amplifies the target gene exponentially, any small variation at the beginning will become unacceptable at the end. Thus, the microplate format is ideal for clinical diagnostics, because multiple samples can be processed simultaneously and equally. Variation can also be quantitated by duplicate or triplicate determinations. Furthermore, the availability of various devices, such as multi-channel pipette, plate dispenser/washer, and liquid handler robot allows us to minimize variation of manipulation.

Acknowledgements

We would like to thank Dr Mike Kobrin (RNAture, Irvine, CA) and Dr Munsok Kim (Hitachi Chemical Co., Ltd., Tokyo, Japan) for their critical review of this manuscript.

References

Alam J, Cook JL (1990) Reporter genes: application to the study of mammalian gene transcription. *Anal Biochem* **188** (2): 245–254.

Aso Y *et al.* (1998) Rapid, stable ambient storage of leukocyte RNA from whole blood. *Clin Chem* **44** (8, Pt 1): 1782–1783.

Chee M *et al.* (1996) Accessing genetic information with high-density DNA arrays. *Science* **274** (5287): 610–614.

Clusel C *et al.* (1993) *Ex vivo* regulation of specific gene expression by nanomolar concentration of double-stranded dumbbell oligonucleotides. *Nucleic Acids Res* **21** (15): 3405–3411.

Duck P *et al.* (1990) Probe amplifier system based on chimeric cycling oligonucleotides. *Biotechniques* **9** (2): 142–148.

Engvall E *et al.* (1971) Enzyme-linked immunosorbent assay. II. Quantitative assay of protein antigen, immunoglobulin G, by means of enzyme-labelled antigen and antibody-coated tubes. *Biochim Biophys Acta* **251** (3): 427–434.

Ferrero M *et al.* (1990) Flow cytometric analysis of DNA content and keratins by using CK7, CK8, CK18, CK19, and KL1 monoclonal antibodies in benign and malignant human breast tumors. *Cytometry* **11** (6): 716–724.

Glisin V *et al.* (1974) Ribonucleic acid isolated by cesium chloride centrifugation. *Biochemistry* **13** (12): 2633–2637.

Hamaguchi Y *et al.* (1998) Direct reverse transcription-PCR on oligo(dT)-immobilized polypropylene microplates after capturing total mRNA from crude cell lysates. *Clin Chem* **44** (11): 2256–2263.

Hyndman D *et al.* (1996) Software to determine optimal oligonucleotide sequences based on hybridization simulation data. *Biotechniques* **20** (6): 1090–1094, 1096–1097.

Jalava T *et al.* (1993) Quantification of hepatitis B virus DNA by competitive amplification and hybridization on microplates. *Biotechniques* **15** (1): 134–139.

Kawasaki ES *et al.* (1988) Diagnosis of chronic myeloid and acute lymphocytic leukemias by detection of leukemia-specific mRNA sequences amplified *in vitro*. *Proc Natl Acad Sci USA* **85** (15): 5698–5702.

Kawasaki ES, Wang AM (1989) Detection of gene expression. In HA Erlich, editor. *PCR Technology: Principles and Applications for DNA Amplification*, pp. 89–97. New York: Stockton Press.

Kievits T *et al.* (1991) NASBA isothermal enzymatic *in vitro* nucleic acid amplification optimized for the diagnosis of HIV-1 infection. *J Virol Methods* **35** (3): 273–286.

Lander ES *et al.* (2001) Initial sequencing and analysis of the human genome. *Nature* **409** (6822): 860–921.

Leone G *et al.* (1998) Molecular beacon probes combined with amplification by NASBA enable homogeneous, real-time detection of RNA. *Nucleic Acids Res* **26** (9): 2150–2155.

Liang P, Pardee AB (1992) Differential display of eukaryotic messenger RNA by means of the polymerase chain reaction. *Science* **257** (5072): 967–971.

Livak KJ *et al.* (1995) Oligonucleotides with fluorescent dyes at opposite ends provide a quenched probe system useful for detecting PCR product and nucleic acid hybridization. *PCR Methods Appl* **4** (6): 357–362.

Lyamichev V *et al.* (1999) Polymorphism identification and quantitative detection of genomic DNA by invasive cleavage of oligonucleotide probes. *Nat Biotechnol* **17** (3): 292–296.

Martel R (2001) Multiplexed assays with ArrayPlates™ applied to mRNA expression profiling. *IBC's 5th Annual Conference on Assay Development,* Abstract, Coronado, CA.

Matsuda K *et al.* (2002) Gene expression analysis from nuclear poly(A) + RNA. *Biotechniques* **32** (5): 1014–1020.

Matsunaga H *et al.* (2002) Application of differential display to identify genes for lung cancer detection in peripheral blood. *Int J Cancer* **100** (5): 592–599.

Melton DA *et al.* (1984) Efficient *in vitro* synthesis of biologically active RNA and RNA hybridization probes from plasmids containing a bacteriophage SP6 promoter. *Nucleic Acids Res* **12** (18): 7035–7056.

Mitsuhashi M (2000) Automation of gene expression analysis. *The Society for Biomolecular Screening: 6th Annual Conference and Exhibition, Vancouver, British Columbia, Canada,* Abstract: 230.

Mitsuhashi M *et al.* (1992) Gene manipulation on plastic plates. *Nature* **357** (6378): 519–520.

Mitsuhashi M *et al.* (1994) Oligonucleotide probe design – a new approach. *Nature* **367** (6465): 759–761.

Miura Y *et al.* (1996) Fluorometric determination of total mRNA with oligo(dT) immobilized on microtiter plates. *Clin Chem* **42** (11): 1758–1764.

Newman PJ *et al.* (1988) Enzymatic amplification of platelet-specific messenger RNA using the polymerase chain reaction. *J Clin Invest* **82** (2): 739–743.

O'Farrell PH (1975) High resolution two-dimensional electrophoresis of proteins. *J Biol Chem* **250** (10): 4007–4021.

Okamoto T *et al.* (1994) Fluorometric nuclear run-on assay with oligonucleotide probe immobilized on plastic plates. *Anal Biochem* **221** (1): 202–204.

Palmiter RD (1974) Magnesium precipitation of ribonucleoprotein complexes. Expedient techniques for the isolation of undergraded polysomes and messenger ribonucleic acid. *Biochemistry* **13** (17): 3606–3615.

Pelkey TJ *et al.* (1996) Molecular and immunological detection of circulating tumor cells and micrometastases from solid tumors. *Clin Chem* **42** (9): 1369–1381.

Perriman RJ, Gerlach WL (1990) Manipulating gene expression by ribozyme technology. *Curr Opin Biotechnol* **1** (1): 86–91.

Ponte P *et al.* (1983) Human actin genes are single copy for alpha-skeletal and alpha-cardiac actin but multicopy for beta- and gamma-cytoskeletal genes: 3′ untranslated regions are isotype specific but are conserved in evolution. *Mol Cell Biol* **3** (10): 1783–1791.

Saiki RK *et al.* (1985) Enzymatic amplification of beta-globin genomic sequences and restriction site analysis for diagnosis of sickle cell anemia. *Science* **230** (4732): 1350–1354.

Sambrook J *et al.* (1989) Extraction, purification and analysis of messenger RNA from eukaryotic cells. In *Molecular Cloning: a Laboratory Manual*, 2nd edn, pp. 7.28–7.52. Cold Spring Harbor, NY: Cold Spring Harbor Laboratory Press.

Schena M *et al.* (1995) Quantitative monitoring of gene expression patterns with a complementary DNA microarray. *Science* **270** (5235): 467–470.

Taniguchi A *et al.* (1994) Competitive RT-PCR ELISA: a rapid, sensitive and non-radioactive method to quantitate cytokine mRNA. *J Immunol Methods* **169** (1): 101–109.

Tominaga K *et al.* (1996) Colorimetric ELISA measurement of specific mRNA on immobilized-oligonucleotide-coated microtiter plates by reverse transcription with biotinylated mononucleotides. *Clin Chem* **42** (11): 1750–1757.

Toulme JJ, Helene C (1988) Antimessenger oligodeoxyribonucleotides: an alternative to antisense RNA for artificial regulation of gene expression – a review. *Gene* **72** (1–2): 51–58.

Tso JY *et al.* (1985) Isolation and characterization of rat and human glyceraldehyde-3-phosphate dehydrogenase cDNAs: genomic complexity and molecular evolution of the gene. *Nucleic Acids Res* **13** (7): 2485–2502.

Tyagi S, Kramer FR (1996) Molecular beacons: probes that fluoresce upon hybridization. *Nat Biotechnol* **14** (3): 303–308.

Urdea MS *et al.* (1991) Branched DNA amplification multimers for the sensitive, direct detection of human hepatitis viruses. *Nucleic Acids Symp Ser* **24**: 197–200.

Venter JC *et al.* (2001) The sequence of the human genome. *Science* **291** (5507): 1304–1351.

Walker GT *et al.* (1992a) Isothermal *in vitro* amplification of DNA by a restriction enzyme/DNA polymerase system. *Proc Natl Acad Sci USA* **89** (1): 392–396.

Walker GT *et al.* (1992b) Strand displacement amplification – an isothermal, *in vitro* DNA amplification technique. *Nucleic Acids Res* **20** (7): 1691–1696.

Walker GT *et al.* (1996) Strand displacement amplification (SDA) and transient-state fluorescence polarization detection of Mycobacterium tuberculosis DNA. *Clin Chem* **42** (1): 9–13.

Wang AM *et al.* (1989) Quantitation of mRNA by the polymerase chain reaction. *Proc Natl Acad Sci USA* **86** (24): 9717–9721.

Wittwer CT *et al.* (1997a) The LightCycler: a microvolume multisample fluorimeter with rapid temperature control. *Biotechniques* **22** (1): 176–181.

Wittwer CT *et al.* (1997b) Continuous fluorescence monitoring of rapid cycle DNA amplification. *Biotechniques* **22** (1): 130–131, 134–138.

Yamazaki T *et al.* (1993) Evaluation of DNA-DNA hybridization method for identification of mycobacteria using a colorimetric microplate kit. *Kekkaku* **68** (1): 5–11.

Zinn K *et al.* (1983) Identification of two distinct regulatory regions adjacent to the human beta-interferon gene. *Cell* **34** (3): 865–879.

A Generalized 96-Well ELISA-Based Assay for Quantitative and Qualitative Monitoring of Cellular Events

7

Laurence Shumway and Lawrence M. Schwartz

7.1 Introduction

One of the key goals in modern biology is to understand, in molecular terms, what a cell 'thinks' about. Subtle alterations in either the pattern of gene expression or post-translational protein modification can have dramatic effects on physiological status of cells. Identifying these changes provides investigators with insights into the basic biology of cellular decision-making. In addition, it can provide researchers with molecular targets for the identification of novel drugs suitable for therapeutic intervention. The use of high-throughput screens to identify altered cellular responses has become a major focus in the pharmaceutical industry.

A variety of biochemical tools have been developed to facilitate the analysis of gene expression at the nucleic acid and protein levels. Not surprisingly, there are advantages and limitations associated with each of these methods. At the nucleic acid level, oligonucleotide 'gene chips' allow simultaneous measurement of hundreds to thousands of transcripts in a single experiment (reviewed in Lockhart & Winzeler, 2000). Instead of assuming that the investigator knows which gene(s) may be relevant to a particular biological process (hypothesis testing), the breadth of these almost genome-wide screens removes target bias and provides a molecular snapshot of the cells' current genetic thoughts (hypothesis generating).

Despite the power of gene chip array analyses, there are several limitations that restrict their universal use. First, these methods involve examining gene expression in the aggregate from a large population of cells, an ensemble measurement, rather than at the single cell level. This may seem like a minor point, but finding that a transcript increases by 10% could mean that every cell had a modest increase in the level of the mRNA, or instead that a small sub-population displayed an exponential increase. The heterogeneous nature of cell populations, where even monoclonal lines will have cells at different stages of the cell cycle or differentiation, means that

Microarrays & Microplates: Applications in Biomedical Sciences, Shu Ye and Ian N.M. Day

ensemble measurements cannot capture this important information for understanding biological responses. Second, the high cost of both the arrays themselves and the equipment needed to process and evaluate them precludes their routine use in screens and reduces the number of replicate measurements made. Typically, conclusions drawn from array studies are based on one or at best three replicate arrays, and inferences based on such small sample size may be weak. Finally, although mRNA levels are important measures of gene expression and indeed may often correlate with protein levels (Futcher *et al.*, 1999), it remains that proteins generally facilitate biological effects and it is their levels that are of paramount interest to researchers. By examining cellular events at the mRNA level, essential post-translational modifications, or altered sub-cellular localization cannot be evaluated.

The more heterogeneous nature of protein chemistry has meant that the development of general-use methods for the global analysis of protein expression has been slower than for techniques to analyse changes at the nucleic acid level. Methods such as coupled 2D chromatographic fractionation and mass spectrometry (Washburn *et al.*, 2001) offer some hope, but are poorly developed and far beyond the means of all but a handful of laboratories.

Fortunately, a variety of methods are available for monitoring the abundance, distribution and post-translational modification of specific proteins in cells. All of these methods rely on the availability of immunoglobulins that preferentially bind, with high affinity, particular antigens. Antisera have been generated that can discriminate between a protein's post-translationally modified isoforms (e.g. addition or removal of inorganic phosphate, sugar, lipid or peptide moieties). Using these antisera, one can determine the abundance of the form of the protein of immediate interest, which has been a great boon for the field of signal transduction.

These antisera can be bound to the epitope of interest either *in situ* via immunochemistry (IC; histological, IHC or cytological, ICC) or when the protein has been immobilized on a support, such as Western blotting. In both cases, exposed antigens are detected when they bind the appropriate primary (specific) antibody. These are detected in turn by the binding of a secondary antibody directed against the species-specific invariant (Fc) region of the primary antibody. The covalent addition of a fluorescent or enzymatic tag on the secondary antibody allows detection of the complex. For example, a horseradish peroxidase tag on the secondary antibody catalyses the conversion of colorless 3,3′-diaminobenzidine (DAB) to an insoluble brown precipitate.

Both Western blotting and IC offer specific advantages and disadvantages for monitoring changes in protein expression. Western blotting is the more quantitative method and is used to determine both the size and relative abundance of the target protein. A population of cells is extracted in a lysis buffer, the proteins fractionated on a porous medium, typically acrylamide, and the proteins are transferred and immobilized on a membrane. The antigen is then detected following incubation with the appropriate primary and secondary antisera. Because it is laborious, this technique is a tedious and impractical primary screening tool and is best used for secondary characterization. In situations where Western blotting is critical for the investigation, some pre-sorting of samples, to minimize the number that must be so processed, is essential.

IC offers a different array of advantages and limitations relative to Western blotting. It is well suited for 96-well format assays, thus allowing large numbers of samples to be processed rapidly. In addition, it can provide information on a cell-by-cell basis. The chief limitation of IC is that it is not very quantitative. Because the commonly used chromagen, DAB, precipitates at the site of the secondary antiserum, cells display punctate staining. Researchers visually examine plates under a microscope and manually assess the level of staining, a process that is both time-consuming and highly subjective. To obtain a more quantitative measurement, examiners often assign various numbers of '+' and '−' to reflect relative levels of staining in the optical field.

Given the demands of modern cell biology, the large number and variety of target molecules to be examined, the requirements for speed, efficiency, low cost and ease of quantification, it is clear that no one assay can satisfy all desired criteria. The hybrid method described in this chapter is designed to provide the quantification of Western blotting with the cellular resolution of IC. As such, it is suitable as a stand alone assay or as a screening tool to reduce the size of the test population of drugs or treatments to a more manageable level. Once a subset of suitable candidates has been identified, more costly or time-consuming methods can be employed for subsequent analysis. For example, it could be used to evaluate a large collection of low molecular mass compounds to identify the subset that alters the level or modification state of a specific target protein within cells. Once this screen has narrowed down the number of credible samples to manageable numbers, more traditional methods such as Western blotting can be employed for subsequent analysis.

7.2 The 96-well *in situ* ELISA assay

The method we present is basically a 96-well *in situ* ELISA assay performed with cultured cells (outlined in *Protocol 7.1* and *Figure 7.1*). An antiserum with the desired specificity is used to label fixed cells. Instead of using DAB as a chromagen, detection of the (poly)horseradish peroxidase-labelled secondary antibody is accomplished with the substrate 2,2-azino-di(3-ethylbenzthiazoline-6-sulfonic acid) (ABTS) (Childs & Bardsley, 1975) as the chromagen (Shumway & Schwartz, 2001). In contrast to DAB, which forms an insoluble precipitate following oxidation, ABTS produces a bright green, water-soluble product that is uniformly distributed within the medium. The oxidized product can be detected on a standard microtitre plate reader at its maximal absorbency wavelength of 410 nm. The measured optical density is proportional to the ensemble average level of the antigen within the population of cells (an example is provided in *Figure 7.2*). As a result, optical density measurements can provide quantitative information about the level of the target antigen within the whole population of cells, delivering some of the virtues seen with Western blot analysis.

With the addition of a subsequent staining step, the same populations of cells can be examined visually to determine the abundance and sub-cellular localization of the antigen under investigation, thus providing the virtues of standard IC. The robustness of the peroxidase enzyme attached to the secondary antiserum allows researchers to wash cells after ABTS colour development and then re-probe them with DAB in order to qualitatively

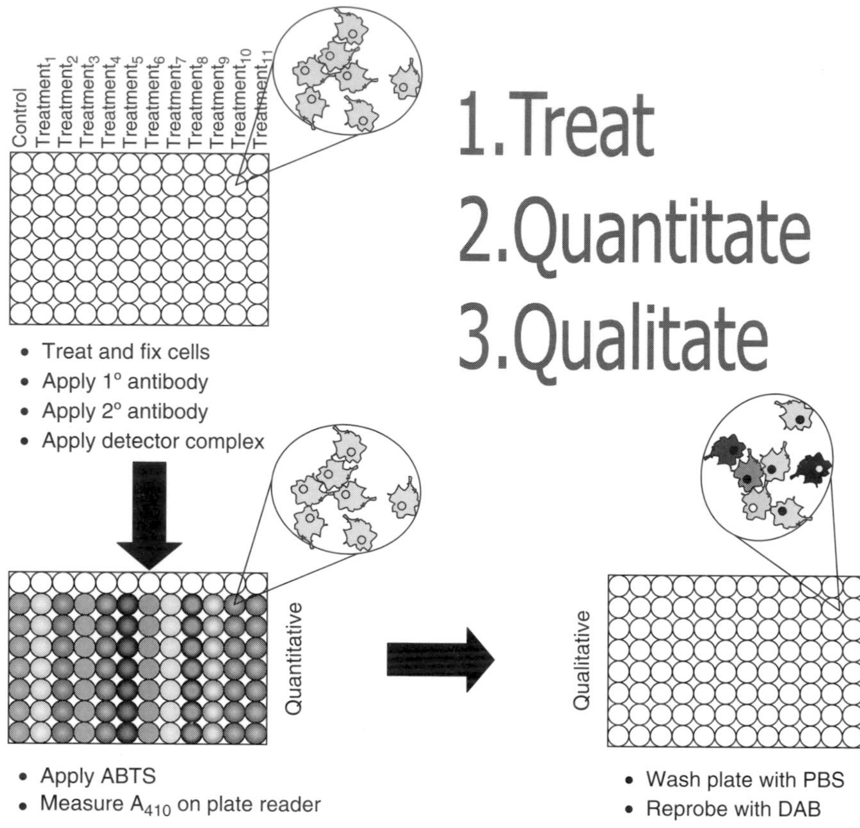

Figure 7.1

The principle steps of the basic assay. A single cell type is seeded across the plate at uniform density. Treatments are applied to columns of cells. Cells are washed, fixed and incubated with epitope-specific antisera for detection of the antigen. ABTS application is the first chromatographic detection step and permits quantitative data gathering via a microplate reader. In the final step, cells are washed and reprobed with a precipitating chromagen such as DAB to localize the antigen for qualitative data gathering. Note that the first well of each treatment (column) is not reacted with primary antibody since it serves as a negative control.

evaluate expression at the single cell level with little loss of sensitivity (*Figure 7.3*).

7.3 Technical considerations

Obviously, the value of this or any other immunoglobulin-based assay is only as good as the antiserum used. If the antiserum employed does not discriminate between several conformational states of the target protein, then assay results will be hard to interpret. For example, if one is looking for agents that alter the initiation of apoptosis by influencing the autocatalysis of pro-caspase-3 into catalytically-active caspase-3, then the antiserum used needs to be able to discern between these two forms of the protease. A large number of manufacturers sell antisera with great

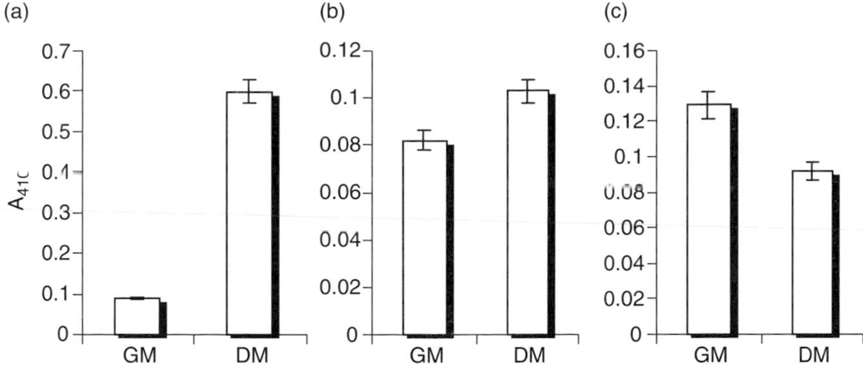

Figure 7.2

Examples of the quantitative data obtained with the assay. C_2C_{12} myoblasts make one of three choices when they are deprived of trophic support by transfer from growth medium (GM) to differentiation medium (DM). Some cells activate both survival and differentiation programmes and fuse to form multinucleated myotubes. As part of the differentiation programme, they dramatically up-regulate the expression of muscle-specific genes, such as myosin heavy chain (MHC) (a). Other cells activate survival programmes to become satellite cells, but do not induce MHC expression. The last population of cells fails to initiate either programme and undergoes apoptosis. This latter process is mediated in part by the cleavage-induced activation of the aspartic acid protease, caspase-3 (b). The Sug-1/p45 26S proteasome subunit is constitutively expressed and can be used to normalize the data for potential changes in cell number that accompany apoptosis (c). (Data from Shumway & Schwartz, 2001; data are given as mean ± standard error for eight replicate wells per group.)

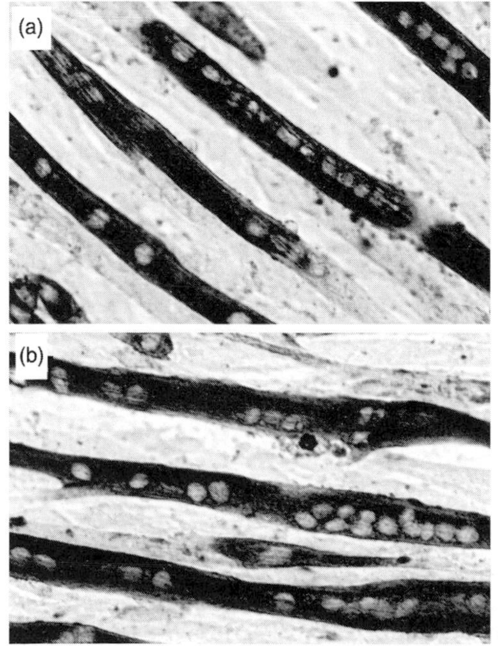

Figure 7.3

Examples of the qualitative visual data obtained with the assay. C_2C_{12} myoblasts were induced to differentiate *in vitro* and stained with a monoclonal antibody directed against myosin heavy chain. Cells were then processed as described in the text and stained with either DAB alone (a) or with ABST followed by DAB (b). Photography and processing were performed identically for both groups of cells. Scale: myotube diameters are ≈30 μm. (Data from Shumway & Schwartz, 2001.)

specificity for the conformation or modification state of proteins that is of biological interest. Careful evaluation of the manufacturer's claims and independent testing in the investigator's laboratory will enhance the likelihood that valuable data will be obtained from the assay.

A second consideration is the choice of fixative to employ in the assay. The conformation of putative target molecules will be different following the use of precipitating fixatives such as acetone, relative to that obtained with cross-linking fixatives such as formaldehyde. Commercially available antisera typically come with recommendations for fixation, and these should be used as a starting point for optimizing the assay. If the antibody is a 'homemade' antiserum, then more work must be performed to determine the optimal fixation and incubation conditions (i.e. temperature and time). Detailed recommendations for performing IC can be found in a number of guidebooks and on the web (i.e. Polak *et al.*, 1997 and http://kpl-nts2.kpl.com/support/immun/protocols.htm). Because each antiserum has its own unique biochemical properties, it is not possible to provide a 'one size fits all' protocol for fixation.

Once the appropriate antiserum and fixative conditions have been determined, it is worth taking some time to ensure that the resulting signal-to-noise ratio is optimal. All the reagents used in the assay are delivered to the cells in a carrier solution, a phosphate-buffered saline (see below). High background signals may arise from the non-specific binding of either the primary or secondary antisera, or the detection complex. To minimize this artefact, 10% serum from the same species that was used to generate the secondary antisera can be added to the carrier solution. This often is quite effective at reducing this non-specific background. Stringency can be further improved by the addition of low concentrations of non-ionic detergents, such as Tween-20 (PBST) or by altering incubation temperatures. From our experience, PBST and serum are sufficient to block most non-specific binding.

Another factor that can give rise to spurious signals is the presence of either endogenous peroxidases or biotin. To determine if these endogenous compounds or other non-specificity interfere with the experiment, perform the entire assay in the absence of the primary antiserum. The implementation of such negative controls is discussed below. The failure to detect a significant signal under these conditions suggests that background signal is unlikely to present problems. If the background is high, then additional steps must be taken.

Some peroxidases are quite hearty and retain their enzymatic activity even after fixation. Endogenous peroxidase activity is particularly troublesome in an assay such as this, as peroxidase catalyses the colorimetric reactions. This background activity can be 'killed' by choosing a different fixative or by treating the tissue with hydrogen peroxide prior to the addition of the horseradish peroxidase. Some protocols suggest incubating cells in 0.6% H_2O_2 in PBS for 30 min in the dark before the addition of the primary antiserum. Treatment to destroy endogenous peroxidase can slightly damage samples, particularly cell cultures, and therefore researchers might find it advantageous to carefully select a fixative that ablates peroxidase activity to avoid this complication.

A separate but related issue is the presence of endogenous biotin, which tightly binds the strepavidin-conjugated horseradish peroxidase, leading also to false-positive signals. Some cells, such as adipose tissue, have high endogenous levels of biotin which is used in lipid synthesis. In these cases, biotin-blocking reagents can be added to the carrier solution prior to the addition of the biotinylated secondary antibody. In our experience with tissue culture cells, endogenous biotin has not proven to be a problem.

After controlling for issues related to cell physiology, controls are still needed to determine the specificity of a particular signal. For example, consider the experimental design outlined in *Figure 7.4*. For a given treatment, the first well should not receive the primary antibody and therefore can serve as a 'negative control'. As mentioned earlier, this step permits the researcher to identify spurious signals. Following the detection step, these negative controls can be evaluated visually or statistically to determine if any adjustments to the subsequent analysis are required. If all wells display comparable negative staining, nothing further is required. If these cells do display an elevated background, the optimization steps described above should be reconsidered. If uneven background levels persist, another possibility is to add an entire column of negative controls, subtract their mean from their corresponding treatment row's mean and sum their variances. The resulting statistic is the mean specific signal of interest. In practice, we typically find that background signals are not a problem.

Once background signals have been adequately addressed, the next issue is to normalize signals between treatments. A researcher wishes to draw conclusions based on observed signal differences that are *de facto* due to treatment differences. Differences in the apparent level of an antigen between two test groups could reflect true differences in per cell expression or differences in cell density arising from errors in initial seeding or different rates of mitosis or apoptosis. Independent of the cause, this potential problem needs rectification if valid inferences are to be drawn. One easy solution, illustrated in *Figure 7.4*, is to employ a duplicate column of cells within each plate for independent staining with an antibody against a constitutively expressed 'house-keeping' protein, whose per cell average level is known *a priori* to be identical between control and experimental lines. Alternatively, if the plate reader can detect fluorescent probes, immunochemical staining can be normalized to the DNA content of the same set of wells using a DNA intercalating dye such as DAPI or propidium iodide. Whichever method is chosen, the next step is to determine if there are significant differences in the density of cells in each well. Simple methods for this comparison are discussed in the Statistical Treatment section below. Basically, the levels of the protein of interest can be normalized to that of the housekeeping protein or to DNA content in order to obtain an accurate measure for comparison between treatments if there are apparent inequities in cellular density. Although this approach is often not required, we have found this normalization step to be useful in situations where multiple independent cell lines are used in an assay, as there is always some variability in seeding density with different parental plates of cells. In assays where a single cell line is seeded into every well, such as screening small compounds on a uniform target population, this normalization step is probably not needed.

7.4 Suggested statistical treatment

There are a variety of approaches that can be employed for the statistical analysis of data obtained with this assay. A few considerations are provided here (and outlined in *Figure 7.4*). These should serve as only a rough, general

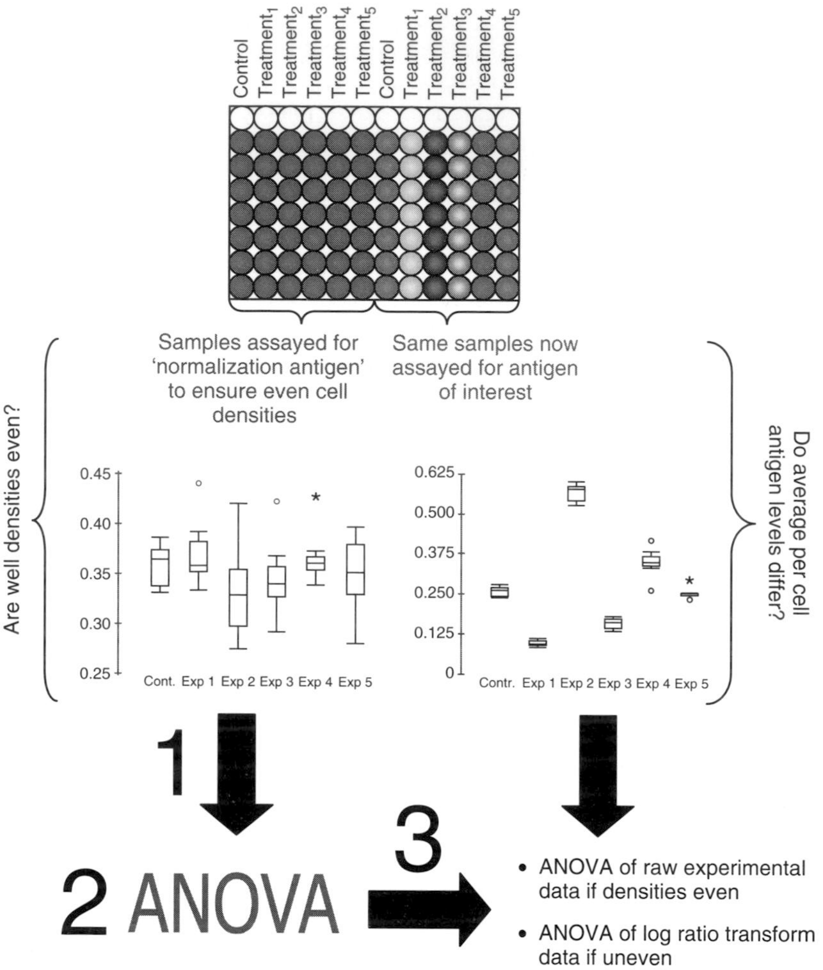

Figure 7.4

Suggested design and checks for signal normalization. For each treatment, two identical sets of wells are seeded. The first is processed for a constitutively expressed 'housekeeping' (control) protein. If one is normalizing to DNA content, this may be done in the same set of wells and only one set will be needed. The second set of wells is analysed for the antigen of interest (experimental). The normalization molecule is one known to vary by a constant proportion across all treatments to the cell density. This set of treatment means (therefore cell densities) is analysed by single-factor ANOVA to ensure even plating. If the normalization data suggest even densities, the experimental data may be analysed, untransformed, by ANOVA then, if warranted, by Tukey. If disparity in cell densities is apparent, experimental data should first be transformed by taking the ratio to the control molecule and taking the logarithm. Boxplots of randomly generated data as examples of normalization (left) and experimental (right) staining are shown.

guide, as different experimental designs may necessitate employing different experimental and statistical approaches.

Our primary use of this method has been to examine the roles of specific genes that regulate differentiative decisions in cells. For example, we have been examining the role of novel regulatory genes that mediate the decision to differentiate or die in murine myoblast (C_2C_{12}) cells that are exposed to differentiation medium *in vitro* (see *Figures 7.2* and *7.3*). We transfect these genes (the treatment) into myoblasts and then measure the expression of downstream markers of myogenesis and apoptosis. To help normalize the signals, we assay a parallel group of cells for a housekeeping gene. We typically use antisera against the 26S proteasome subunit Trip-1/sug-1 26S, as this protein appears to be constitutively expressed throughout development and differentiation in these myoblasts (Hu *et al.*, unpublished). Most investigators, however, will probably be better served by choosing actin as the housekeeping reference as there are good monoclonal and polyclonal antisera are available and in most situations, actin is constitutively expressed. (This is not the case for myoblasts, as actin expression is induced with myotube formation.)

The statistical procedure that we have most commonly employed is a single factor analysis of variance (ANOVA). To proceed accordingly, the following criteria must be met: (i) antigen levels within a given treatment (column) must be independently and normally distributed; and (ii) equality of variances, the variance among wells within a treatment should be equal across all treatments. A normal quantile plot assesses the normality of the data. As for variance, if the largest and smallest variances in the set differ by no more than approximately twofold, it is generally safe to assume all variances are equal. We have found that these criteria are typically met.

Normalizing a signal of interest to a baseline (cell number) is commonly achieved by a log ratio transformation of the signal of interest (taking the log of the ratio of the level of interest to either your housekeeping gene or nucleic acid level). The log transform of the ratio is for convenience and investigators are encouraged to explore other transforms, especially those for which there is a rational foundation. The resulting statistics are usually normally distributed with approximately equal variances, and can therefore be examined by single-factor ANOVA. By convention, 'I' will be the number of treatments (the number of different columns of treated cells), while 'J' is the number of observations in each treatment (the number of wells per column receiving identical treatment). In our example, the number of observations will be seven, if treatments are applied column-wise and the first well of each column is used for negative control. The statistic of interest in the ANOVA is the *F*-ratio, with $I - 1$ numerator df and $I(J - 1)$ denominator df. If the *F*-statistic deviates significantly (consult a readily available table) from 1, it suggests that there are differences amongst the true treatment means that are not attributable to chance. However, this analysis does not say *which* means differ and the method of Tukey is useful for addressing this question. The procedure begins by ordering your means numerically. Determine the width of an appropriate confidence interval (of level $[1 - \alpha]$) by a Studentized range distribution (for which tables exist) on I numerator df and $I(J - 1)$ denominator df times the sqrt(MSError/J) (CI width $= Q_{\alpha,I,I(J-1)}\sqrt{(MSE/J)}$). Means that differ by less than this interval width are underscored together to indicate that they do not differ significantly. Consequently, means not underscored by the same line differ significantly (DeVore, 2000). Post-hoc techniques such as those of Scheffe, can estimate the pair-wise differences of means, ascribe a level of confidence to them and allow an investigator to distil practical significance from statistical significance.

7.5 Additional applications of the assay

As described above, this assay can be employed in any situation in which a suitable antiserum is available to monitor changes in the absolute levels of a given protein, or a series of post-translational modifications, such as cleavage, phosphorylation, or glycosylation. In some experiments, *de novo* expression of an epitope-tagged protein can be determined following the addition of the appropriate transcriptional regulator. For example, in transfection studies, tetracycline-induced expression of a given gene can be quantified without the need for FACscan analysis or Western blotting. In separate studies, this method can be used to monitor changes in cellular physiology not directly related to protein levels. For example, bromodeoxy-uracil (BrdU) can become incorporated into newly synthesized DNA and is used as a measure of mitosis in cells. By using an anti-BrdU serum, the fraction of dividing cells can be quantified without the need to visually inspect fields of cells in tedious manual counting.

In theory, this method can also be extended for the analysis of tissue samples derived from *in vivo* sources. Cryostat or paraffin sections containing tissue can be rehydrated, placed in wells and reacted with the primary antisera of interest. After processing with ABTS, the endogenous level of the antigen can be quantitatively determined. As before, subsequent DAB staining then allows cellular distribution of the antigen to be evaluated. For example, overall levels of apoptotic cell death in the brain following ischaemia can be determined by using an anti-activated caspase-3 antibody (i.e. Cell Signaling Technology, Beverly, MA). Secondary staining with DAB then permits the researcher to determine, by visual inspection, which particular brain regions or cell types are most significantly impacted.

The applications described thus far represent analysis at a specific time following treatment. However, it is possible to extend this analysis to provide temporal information about changes in protein expression. A randomized block design, in which time is the blocking variable, can allow the researcher to distil more subtle treatment differences. Following a specific signal over time by generating a linear model provides a more informative, holistic picture of the process under investigation and presents no new technical challenges beyond those discussed here.

7.6 Conclusion

In summary, the assay described in this chapter is suitable for a wide variety of applications in cell biology. The ability to recover both quantitative and qualitative data easily and quickly makes it attractive as both a stand-alone assay and as a screening tool to reduce the number of samples that need to be analysed by more tedious methods. We have highlighted what we believe to be the most valuable applications and important concerns in its implementation. Researchers using the assay will no doubt refine and expand the method.

Acknowledgements

We thank Justyne Ogdahl for generating *Figure 7.3* and Ms. Joanna Beinhorn from Cell Signaling Technologies for generously providing the activated caspase-3 antibody used in the development of this assay. Supported by grants from the NIH, the Arnold and Mabel Beckman Foundation and the University of Massachusetts Honors College.

References

Childs RE, Bardsley WG (1975) The steady-state kinetics of peroxidase with 2,2′-azino-di-(3-ethyl-benzthiazoline-6-sulphonic acid) as a chromogen. *Biochem J* **145**: 93–103.

DeVore JL (2000) *Probability and Statistics for Engineering and the Sciences*, 5th edition. Duxbury: Thomas Learning.

Futcher B *et al.* (1999) A sampling of the yeast proteome. *Mol Cell Biol* **19**: 7357–7368.

Lockhart DJ, Winzeler EA (2000) Genomic, gene expression and DNA arrays. *Nature* **405**: 827–836.

Polak JM *et al.* (1997) *Introduction to Immunocytochemistry*, 2nd edition. New York: Springer.

Shumway J, Schwartz LM (2001) A generalized 96-well format for quantitative and qualitative monitoring of altered protein expression and post-translational modification in cells. *BioTechniques* **31**: 996–999.

Washburn M *et al.* (2001) Large scale analysis of the yeast proteome by multidimensional protein identification technology. *Nat Biotechnol* **19**: 242–247.

Protocol 7.1: The 96-well *in situ* ELISA assay

METHOD

1. Seed cells in sterile, flat-bottomed 96-well tissue culture plate at desired density. (We typically use 8000 cells per well to achieve 80% confluency for murine C_2C_{12} myoblasts.)
2. After treatment, remove media, wash once with phosphate-buffered saline (PBS), then fix cells. (Note: antigen presentation may be altered with different fixatives and preliminary experiments should be employed to determine which treatment is optimal for the antiserum employed, see below.) For example, we have found that a 1 h fixation in 3% paraformaldehyde at 4°C has worked well with most of the antisera we employ with our murine C_2C_{12} myoblasts.
3. Wash well three times with PBS containing 0.1% Tween-20 (PBST).
4. Block for 30 min in PBST with 10% normal serum of the species used to generate the secondary antiserum (which is referred to as 'carrier solution').
5. Incubate overnight at 4°C with primary antibody diluted in carrier solution.
6. Wash three times with PBST.
7. Incubate for one hour with the appropriate biotinylated secondary antiserum diluted in carrier solution.
8. Wash three times with PBS.
9. (Optional) Treat cells with 0.6% H_2O_2 in PBS for 30 min in the dark at room temperature to eliminate endogenous peroxidase activity.
10. Wash three times with PBST.
11. Incubate for 40 min with avidin–biotinylated horseradish peroxidase, which consists of a 1 : 1 mixture of reagents 'A' and 'B' from Vectastain Elite ABC kit (Vector Laboratories, Burlingame, CA) diluted 1 : 400 in PBST.
12. Wash three times in PBS.
13. Incubate in the dark for 20 min in 1 × ABTS reagent (Vector Laboratories).
14. Read plates on a microtitre plate reader set to measure absorbency at 410 nm.

15. Wash cells once in PBS.
16. Incubate (in the dark) at room temperature in 1 × Sigmafast DAB/urea H_2O_2 reagent (Sigma, St. Louis, MO) until the desired intensity is observed.
17. Remove DAB, wash once in PBS and store in PBS at 4°C.
18. Unless otherwise noted, incubations may be at either room temperature or 37°C. Delivery of solutions to the wells is most easily accomplished with an 8- or 12-tip pipette. Removing solutions can be done a number of ways. If cells are reasonably adherent, inversion and gentle shaking of the plate is fast, easy and doesn't damage cells.

Electrophoresis in Microplate Formats

8

I. N. M. Day, M. A. Al-Dahmesh, K. K. Alharbi, X. Chen, R. H. Ganderton, T. R. Gaunt, L. J. Hinks, S. D. O'Dell, E. Spanakis, P. J. R. Day, M. A. Suchard, B. B. Zhang and M. R. James

8.1 Introduction

The well-known industry standard 96-well microplate has evolved to form the basis of many high-throughput approaches in the biosciences over the past 30 years. However, microplates are only permissive of liquid-phase reactions or solid-phase (binding) separations, whereas electrophoresis can derive information about parameters such as size, shape and charge of molecular moieties, as well as acting as a highly resolving separation approach for complex mixtures. In 1994, we described a generalized system, 'microplate array diagonal gel electrophoresis' (MADGE) that has a number of advantages for laboratory molecular genetic analysis of population samples (Day & Humphries, 1994a, b). This system contests the traditional wisdom that electrophoresis should be avoided and liquid-phase analyses developed where high-throughput applications are contemplated. The system combines liquid-phase microplate compatibility (laboratory transfers and informatics), convenient set up and use of polyacrylamide gel electrophoresis (PAGE; which has higher resolution than agarose post-PCR). Short track length/short run time, compactness and scalability with hand-sized robust slab gels accessible for direct human interaction, that is mini not micro scale, are additional features. Start-up costs are minimal, so 'third world' laboratories can also apply the system. Essentially, any analysis which can be reduced to short tracks (e.g. <3 cm) gains the throughput advantages. One technician can process 10–100 gels (1000–30 000 tracks) per day according to the application and the MADGE implementation in use. MADGE is applicable to many categories of DNA analysis (originally PCR-RFLP and allele-specific PCR analyses, but more recently for minisatellite and microsatellite sizing and the identification of unknown mutations) and potentially to many other categories of biomolecule (protein, etc.) and we can only speculate on future utilization. Complete genome mapping (simple but very high-throughput PCR checking gels from radiation hybrid cell lines) has also been implemented (Watanabe *et al.*, 1999).

In 1807, Ferdinand Frederic Reuss observed under a microscope the migration of colloidal particles in an electric field, perhaps the first electrophoretic separation. During the early 1970s both agarose and polyacrylamide

Microarrays & Microplates: Applications in Biomedical Sciences, Shu Ye and Ian N.M. Day

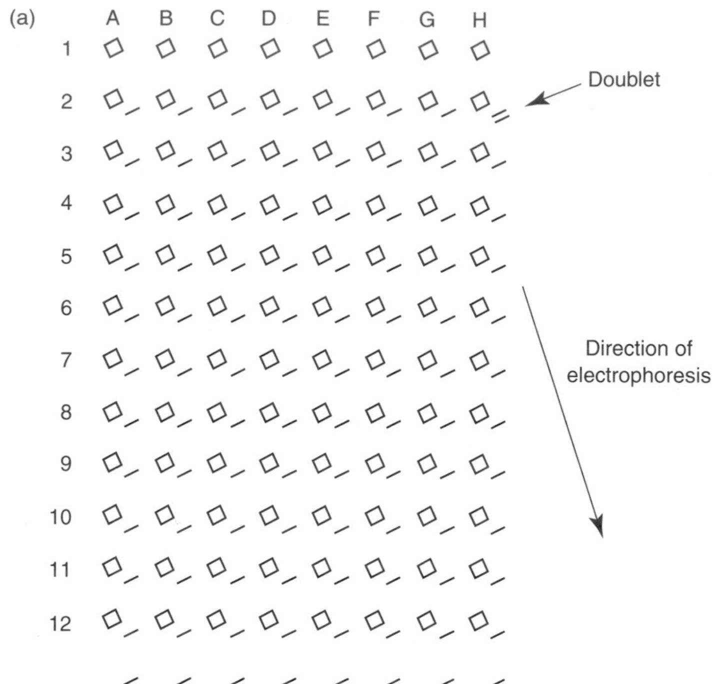

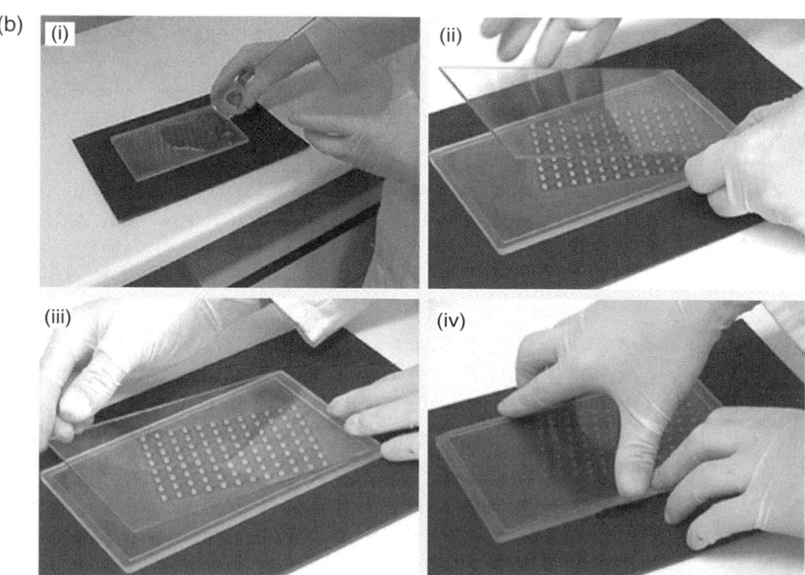

Figure 8.1

Microplate array diagonal gel electrophoresis (MADGE). (a) Schematic diagram of a microplate array diagonal gel electrophoresis (MADGE) image. The squares represent the wells and the lines represent the bands in the lanes. Using ethidium bromide staining, the wells are only visible as a dark image in a dark background, whereas the bands stand out brightly in a regular array. Although seemingly complex, the human eye and brain are efficient at pattern recognition. For example, here, only track H1 contains a doublet. The set-up is such that well A1 is placed nearest the cathode.

electrophoresis gel usage evolved to slab gel formats (Studier, 1973; Sugden *et al.*, 1975) much as they are still found in the laboratory today. Throughput requirements were not an issue at that time. Indeed, slabs were an advance from tube gels. For both tubes and modern capillaries, each 'lane' is separate from any other and cross-referencing requires reliance either on an external frame of reference (e.g. time of elution) or an internal frame (e.g. co-electrophoresed mobility standards). Slabs give convenient alignment of tracks. MADGE, in the process of achieving microplate compatibility, uses 96 (higher density in newer formats) track origins (wells) with 8 × 12 array locations at the 9 mm pitch identical to microplates devised 30 years ago by Dynatech and now a highly established industry standard for liquid-phase operations. The long axis of the array is at an acute angle, for example 18.4°, relative to the direction of electrophoresis (e.g. *Figure 8.1a*) so that a track length of 25–30 mm is available. It should be noted that each track is specifically isolated (like a set of tube gels or capillaries) but the tracks are in a slab. The two-piece kit dimensional gel former and piece of backing glass or plastic (*Figure 8.1b*) for the gel, both excludes air and gives a perfectly flat open face to the resultant gel, simplifying the production of agarose, PAGE, and other gel matrix or composite gels. The combination of features in MADGE gel design gains substantial advantages (as described above and below) but the features of this system can be used in isolation, for example the simplified production of open faced horizontal PAGE gels (Day & Humphries, 1994a) but with standard row(s) of wells.

High throughput DNA studies (at present almost all derived from PCR) demand economy, sample parallelism, convenience of set up and accessibility to small as well as large laboratories, goals inadequately met by existing approaches. For example, systematic association studies demand the analysis of many thousands of candidate variations, commonly single nucleotide variations. Power studies (Risch & Merikangas, 1996) indicate that for common disease traits in which individual genotypes make smaller percentage contributions, cohorts, trios or other architectures comprising hundreds or thousands of individuals or nuclear families will be essential. Such studies can be prohibitively expensive both in phenotyping and genotyping, and therefore the major hurdles to overcome are in cost reduction. These needs have driven the laboratory technology inventions described here and we have also extended these developments into microsatellite marker sizing and to the identification of unknown mutations. Electrophoresis is well known and available to all users, in contrast to the expensive hardware implicit in homogeneous techniques, and a range of well-established methodologies are already centred around electrophoresis [amplification refractory mutation system (ARMS), restriction analysis, heteroduplex analysis, denaturing gradient gel electrophoresis (DGGE), single strand conformation polymorphism

Figure 8.1 *(Continued)*

(b) Preparation of a MADGE gel, original and simplest format. The plastic former is laid horizontal with the teeth facing upward. Acrylamide gel mix is poured (i) into the 'swimming pool' which contains the teeth. A sticky silane-coated glass plate is laid over (ii–iv). Once the gel has set (1–5 min if higher concentrations of polymerization agents are used), the glass plate is prised off, bearing an open-faced microplate-compatible 96-well PAGE gel suitable for submersible or semi-dry use.

technique (SSCP), protein electrophoresis]. However, gel preparation, long track lengths, incompatibility with industry standard microplates, vertical format polyacrylamide gels and a host of other general inconveniences, make electrophoresis an unattractive (although used) option to underpin laboratory studies of genetic diversity within populations. We have aimed to eliminate these disadvantages. Several representative developments are described in this chapter.

1. 96-Well MADGE for PCR checking and for assays of known single nucleotide variants (Day & Humpries, 1994a, b; Bolla *et al.*, 1995; Day *et al.*, 1995a, b; O'Dell *et al.*, 1995, 1996). Implementation (AutoMADGE) for gel reuse and more automated software analysis for very high-throughput PCR checking (Watanabe *et al.*, 1999).
2. 192-Well MADGE (O'Dell *et al.*, 2000a), 384-well MADGE, 768-well MADGE (Day & Wilson, 2001) and combinations with ARMS PCR reactions and software analysis for single nucleotide polymorphism typing (O'Dell *et al.*, 2000a).
3. Higher resolution MADGE, such as for minisatellite (O'Dell *et al.*, 2000b) and tetranucleotide and trinucleotide microsatellite polymorphisms (Chen *et al.*, 2002).
4. Melt-MADGE, for rapid, high-throughput, *de novo* mutation scanning of PCR products (Day *et al.*, 1995a, 1998).
5. Software for MADGE gel image analysis.

There is not space in this chapter to give detailed protocols for all of the developments. A 384-well protocol has been selected, whereas outline detail is given for the principles of each of the other approaches.

8.2 MADGE, 96-well formats

Electrophoresis of DNA has traditionally been performed either in agarose or polyacrylamide gel matrix. Much effort has been directed to improved quality agaroses capable of high resolution, but for small fragments, such as those from PCR and post-PCR digests, polyacrylamide still offers the highest resolution. Although agarose gels can easily be prepared in an open-faced format to gain the conveniences of horizontal electrophoresis, acrylamide does not polymerize in the presence of air and the usual configurations for gel preparation lead to electrophoresis in the vertical dimension.

The original MADGE format (Day & Humphries, 1994a) uses a 2D plastic former in conjunction with one glass plate coated with gamma-methacryloxypropyltrimethoxysilane (Sticky Silane). The plastic former contains a 2 mm deep, 100 × 150 mm rectangular 'swimming pool'. Within the pool, there are 96 2 mm cubic 'teeth' (well formers) in an 8 × 12 array with 9 mm pitch directly compatible with 96-well plates. The array is set on a diagonal of 18.4° relative to the long side of the 'pool', which is parallel to the eventual direction of electrophoresis, giving gel track lengths of 26.5 mm (*Figure 8.1a*). The gel-former is placed horizontally, acrylamide mix poured into the pool and the glass plate overlaid (*Figure 8.1b*). After the gel has set, the glass plate is prized off, bearing its open-faced 96-well gel. The original description of apparatus and gel set up is given in Day and Humphries (1994a). It should be noted that although the system may sound complex, many users have noted that this is the simplest gel system they have ever used. Additionally, the human eye/brain is extremely good at pattern recognition and the small-scale user, once accustomed is not obliged to use image analysis software. A 26.5 mm track length (2 × 4 diagonal array) such as in *Figure 8.1a* or greater such as the 2 × 6 diagonal in *Figure 8.5* (see section higher resolution MADGE), using polyacrylamide gel matrix, is efficient for many post-PCR analyses involving simple band pattern recognition. Because the simplest situation for post-PCR analysis is to load PCR product directly onto the gel, the question arises about the effects of carry-over buffers and salts in the sample solutions on separation in MADGE. Our typical PCRs contain 50 mM KCl and loading 5 μl leads to a reasonable resolution of patterns in which relative mobility differences between bands are 5% or greater. Extremely high salt digests or other less typical solutions may occasionally present a problem. Rather than resort to laborious ethanol precipitations of many microplates of amplicons, or other expensive purification procedures, we have usually been able to 'dilute away' the solute problem by diluting the sample in the gel buffer without any compromise to band detection. If ethidium bromide then turns out to provide insufficient sensitivity, order of magnitude improvements in sensitivity are readily achievable with newer intercalating dyes such as Vistra Green. Indeed, poorer resolution of doublets with <5% mobility differences has often simply been the consequence of band overloading. The thicker (2 mm) gel used with transverse imaging would also offer lower resolution if bands were significantly tilted in orientation from glass toward gel surface. This is no greater a problem than for horizontal submerged agarose gels and generally this has not been a problem at all in our laboratory. Attention to detail of buffers and ions in gel and sample (Kozulic, 1994) may improve results further but for the typical applications described it must be stressed that direct loading of samples usually gives perfectly readable results. Gels are robust, reusable, directly stackable for storage or use, and fully compatible with industry-standard PCR microplates thus minimizing procedural information transfers and enabling direct 96-channel pipetting. However, considerable advantage is gained even with 8 and 12 channel pipettes.

AutoMADGE is the informal laboratory name used to refer to a particular implementation (*Figure 8.2*) of MADGE developed in Oxford for high-throughput genomics projects. The main application was in radiation hybrid mapping in which the presence or absence of a PCR product is scored in a panel of radiation hybrids. MADGE was ideal for these projects because any given project utilized a single radiation hybrid panel that was accommodated on a single microtitre dish and the whole process can be reduced to a highly repetitive operation. Although radiation hybrid mapping is operationally quite simple, high-throughput projects such as the Human Gene Map (Schuler *et al.*, 1996) have historically employed considerable labour and materials. Alternatives such as various homogeneous assay formats or dot blot approaches (Genomatron, Whitehead Centre for Genome Research) tend to be too expensive or cumbersome to optimize for the many tens of thousands of different loci surveyed. We wished to avoid the large teams of people required for pouring, loading and electrophoresis of standard submarine gels. We focused on a system that was very efficient in the preparation of large batches of gels, ideally allowed reuse of the gels and took advantage of the MADGE format for rapid, error-free loading of samples, rapid electrophoresis and analysis. A brief outline of the system is described here with a more complete technical description to be presented elsewhere (Day *et al.*, unpublished).

The standard 96-well MADGE format was used but special custom-made gel carriers (*Figure 8.2*) allowed the central part of the gel, comprising all of the

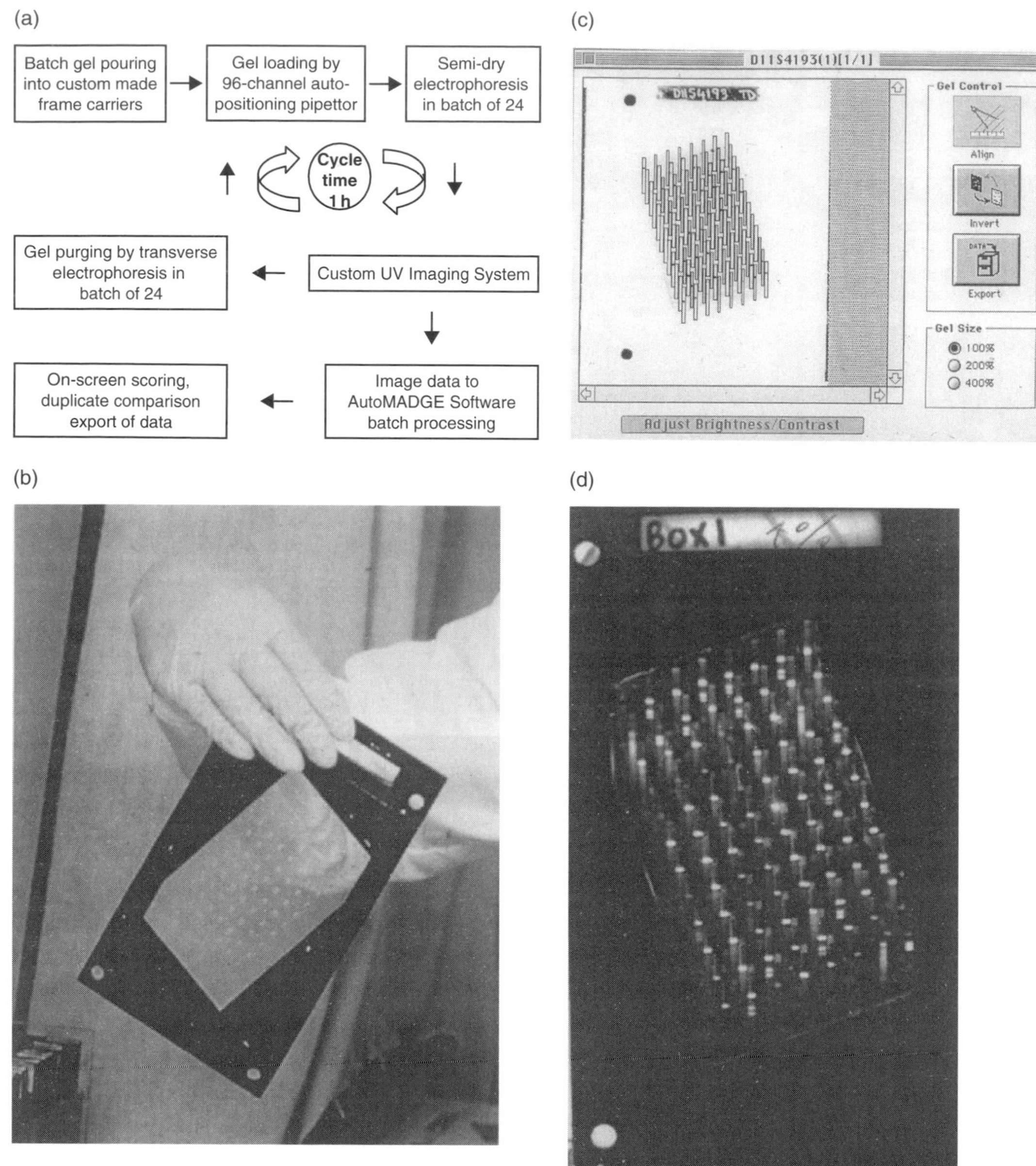

Figure 8.2

AutoMADGE, a very high-throughput 96-well MADGE system employing reusable gels. (a) Schematic of workflow. (b) A gel formatted for multiple reuse. The gel is shown held vertical in its plastic frame – the oblong-shaped cut out corresponds to the suspended area of the gel. (c) A computer screenshot of the software performing the first stage of gel analysis: aligning the gel and demarcation of the 96 lanes prior to redrawing into two rows of 48 lanes. Various tools allow manual intervention and optimization but this step is normally performed automatically to process batches of 24–96 gels. (d) Original gel image of a matrix metalloproteinase-9 PCR-RFLP analysis; (e) gel image (d) after conversion in software to calling/validation screen.

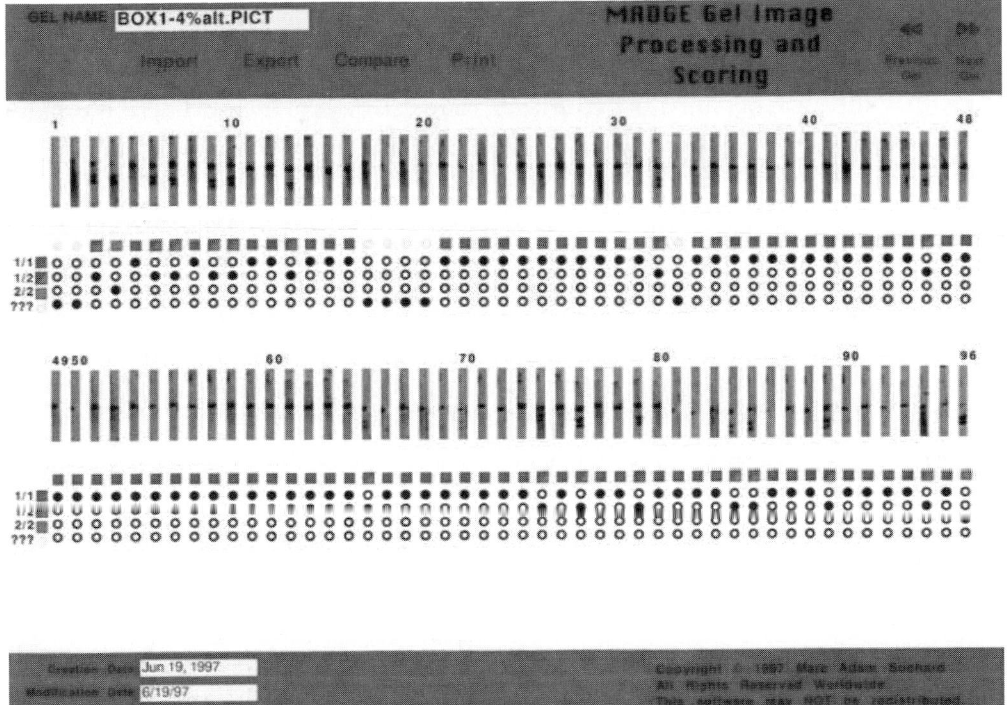

(e)

Figure 8.2 *(Continued)*

sample loading and electrophoresis zone, to be suspended free of any support. Composite PAG and agarose gels are used in which the electrophoresis separation was largely determined by the PAG (generally 3.5–4.5%) and agarose provides the strength to allow suspension. The composite polymer is very tensile and robust. Fortuitously, this also means that only tiny amounts of expensive agarose are used compared with regular, or even similar-dimensioned agarose gels. During most operations such as sample loading, electrophoresis and UV imaging the gel rests on the flat surface of the various apparatus, but the ability to suspend the gel without support and with buffer on upper and lower surfaces allows samples to be electrophoresed out of the gel in the Z dimension, that is, the electric field is applied at right angles to the normal migration position. This operation is not dissimilar to electroblotting and the apparatus consists of vertical plate electrodes at both ends of a tank that holds an extractable rack with 24 MADGE gels held vertically and a high wattage electroblotting power supply is employed. The gels are typically 2 mm thick and, in a transversely applied electric field, even large DNA is purged from the gel in a very short time – most DNA disappears within a few minutes. In this way we have been able to reuse a set of 24 MADGE gels more than 30 times.

Other relatively simple apparatuses to support related and necessary operations were developed. A simple pneumatic 96-channel pipettor was mounted on a stand (in fact a domestic electric drill stand) with vertical axis movement which allowed samples to be transferred from the microtitre plates to the gels in a few seconds and with a precision independent of the operator. The gels were electrophoresed in the 'semi-dry' mode on a flat bed with the underside of the gel resting on a carbon cathode and a stainless steel anode at respective ends. Under quite low field strengths, rapid separation was obtained, actually faster than can be exploited so that it is usually slowed down to ≈15–20 min. A modified gel imaging system employed a permanently on UV light source and a sliding tray to hold two gels in fixed positions that was slid into the UV/camera enclosure. Two gels were imaged simultaneously with software-processing into separate image files. A batch of 24 microtitre plates can be processed in ≈50–55 min, that is, gel loading, electrophoresis, UV imaging and electro-purging of the gel for reuse. The gels are partially destained in this procedure and the rack of 24 purged gels were allowed to rest in electrophoresis buffer for at least 1–2 h before reuse.

The MADGE gel support system also allows sample identification information (name of the experiment/gel, etc.) to be carried with the gel throughout all manipulations and to appear in the final image, making the overall system extremely robust to sample tracking problems. Reference marks are built into the gel carriers and also appear in the image to enable custom software to automatically locate the 96 wells. The software then redraws and deconvolutes the 96 lanes into two rows of 48 side-by-side lanes for easy visual scoring and registration on-screen using a mouse. The software provides for semi-automated comparison of duplicate gels or the same gel independently scored by two operators.

With up to 4 sets of 24 gels in constant rotational use this system has allowed us to perform up to 12 runs in a 24-h period with 2–4 operators. This peak throughput represents 27 648 PCRs having been loaded, electrophoresed and imaged in one day, all without a single gel being poured at the time of the experiment. In practice the system is rarely used close to its maximum throughput due to other bottlenecks, mostly sample preparation and PCR. However, the ability to perform the gel step at these very high rates does allow for more efficient implementation of the different phases of the experiment – for example, one might set up and perform PCR on large numbers of plates over, say, a week and then analyse them all in one or a few days of gel work. It has allowed us to perform significant genome projects either solo (Watanabe *et al.*, 1999) or as consortia (Schuler *et al.*, 1996; Deloukas *et al.*, 1998) without the large teams and corresponding large budgets typical of such endeavours. The system is also very economical on laboratory space in that the gel loading, electrophoresis and UV imaging modules occupy <3 m of bench.

Although the composite gels used in AutoMADGE have somewhat less sharp resolution than pure PAGE gels, they are still superior to high-percentage agarose gels and have been used for polymorphism assays. A version of custom software developed in conjunction with the AutoMADGE system allows on-screen scoring of diallelic polymorphisms (Zhang *et al.*, 1999). An example is shown in *Figure 8.2*.

8.3 192-Well, 384-well, 768-well MADGE, ARMS and SNP typing

8.3.1 192-Well MADGE

For checking one or two bands, much of the usual MADGE track length remains unused. One general approach to SNP is to use allele-specific PCR

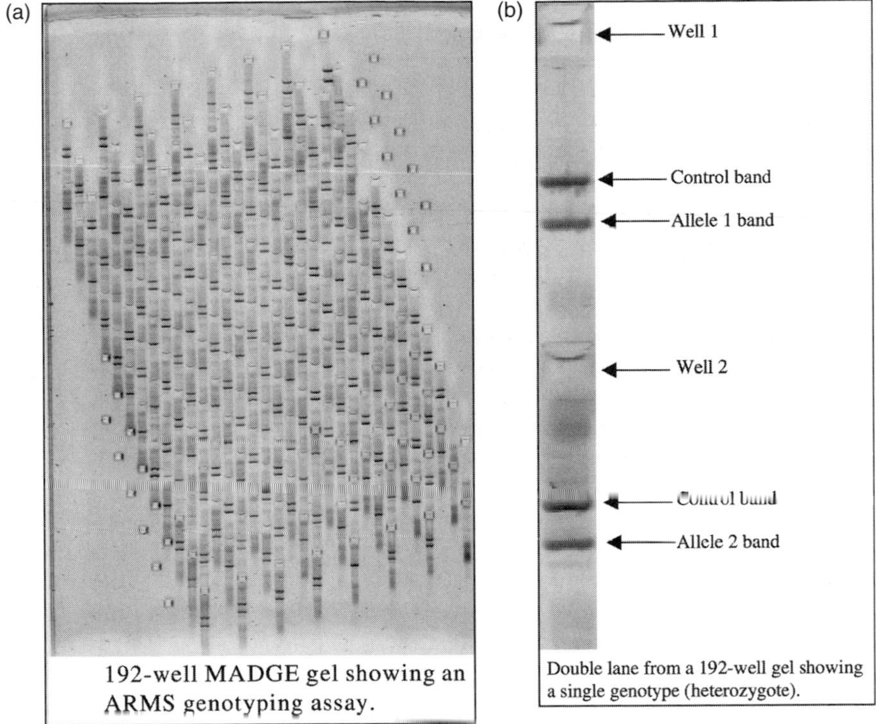

Figure 8.3

192-Well MADGE in conjunction with ARMS analysis of a SNP in 96 samples (2 allele reactions per sample). (a) One panel shows a complete 192-well gel image, the other a 'tandem' track in which the first well is loaded with the PCR assay for one allele in a subject, the second well is loaded with the PCR assay for the second allele in the same subject. (b) The pair of wells are located such that one 'virtual' track is obtained directly upon viewing or during software analysis.

such as 'ARMS' (Newton *et al.*, 1989): one simple robust format uses two separate allele specific amplifications, one testing for each allele, with a control (unlinked) amplicon also included in each reaction as a positive control. In this case, it is easiest to load one gel and recombine the two tracks of information about a given template into one 'virtual' software track. This can be readily achieved by locating a second well half way along the 'original' MADGE track, giving two concatenated tracks of 12.25 mm (for 2 × 4 diagonal MADGE). This is exemplified in *Figure 8.3*.

8.3.2 384-Well MADGE and 768-well MADGE

384-Well microplates have become an established new standard in higher throughput genomics laboratories for clone operations (arrayed libraries, gridding, spotting, storage, etc.) and for PCR. The difficulties in thin-walled plastics manufacture for the latter have been solved. 384-Well microplates have wells in a 16 × 24, 4.5 mm pitch array. A diagonal turn of this array would result in MADGE wells and tracks too narrow (<1 mm) for manual

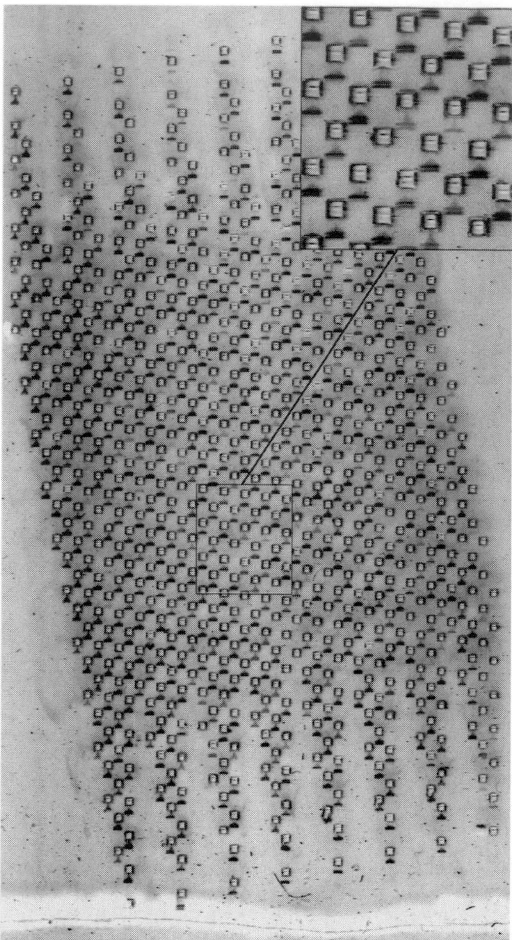

Figure 8.4

768-Well MADGE. In each track of this gel, the presence of one band denotes amplification of control amplicon only, whereas two bands denote positive amplification also for a specific allele in an ARMS PCR test. Track width, real size, is 1.5 mm. Gel set up and loaded by 96-pin passive replicator, exactly as for 384-well gels protocol given in this chapter.

access, but we have devised a 384-well MADGE format which combines four 96-well arrays in a linear (rather than tetradic) overlay. The 1.5 × 1.5 × 1.5 mm wells achieved can be accessed by human hand. However, the array is too dense and confusing and is undesirable for 8- or 12-channel loading, but 96-pin passive or air displacement transfer is not a problem for human hand and eye. We have also derived a 768-well format from 384 (proof of principle in Day & Wilson, 2001, see also *Figure 8.4*) in the same way that the 192-format was derived from the original 96-well format (see above). Protocols for setting up 768-well gels will be similar with those given below for 384-well gels. The 384-well protocol depends on the use of oil-free PCR, 96-pin manual transfers from a 384-well PCR plate to gel, and the use of dry rather than submerged gels to enable use of passive pin transfers: this protocol is described in detail in the Protocols section, see also *Figure 8.7*. The present approach to machining of the gel formers is near its limit because the space between 'teeth' is small, but other arrangements of teeth, or other approaches to gel former production, may enable yet higher densities. However, it is likely that this would force the transition from economy of start up and versatility of the human operator, to expensive hardware configured around a core mini-micro scale format and at this

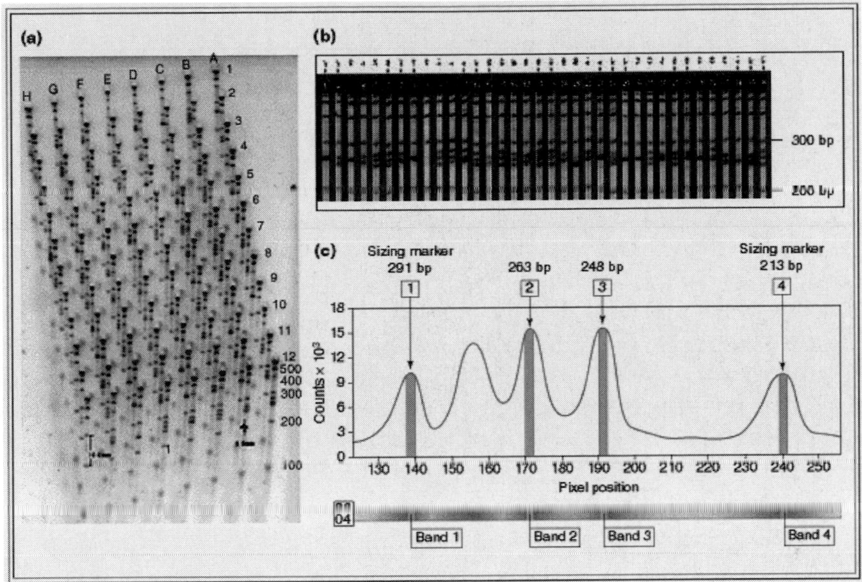

Figure 8.5

High-resolution MADGE analysis of a multiallelic microsatellite polymorphism. Products were run double stranded, so both homoduplexes and heteroduplexes are evident. In the heterozygote, the typical pattern is two resolved homoduplexes of different sizes, and a third lower mobility band in which the pair of heteroduplexes co-migrate. Mobilities are referenced to background mobility markers introduced in the sample loading buffer to every track. A five allele system (15 genotypes) (*HUMTH01*) is shown. (a) Gel image. (b) Phoretix™ software single-row presentation of the 96 MADGE lanes. This layout illustrates the power of MADGE to resolve the different genotypes clearly. (c) Peak intensity profile produced by the Phoretix software. Bands are indicated with arrows; an algorithm is used to assign sizes to the unknown bands relative to the within-track size markers. This example shows *HUMTH01* homoduplex (and heteroduplex) bands bracketed by size markers of 213 and 291 bp.

level, complete departure from industry standard microplate format may be appropriate.

8.4 96-Well high-resolution MADGE

Attention to detail (increased track length on 2 × 6 diagonal instead of 2 × 4; introduction of internal molecular mass markers in every track; sample dilution to avoid salt artefacts in the electrophoresis; use of better resolving acrylamide derivatives such as duracryl; thermostatic control of the electrophoresis) enables resolution to <1.5% mobility differences of bands (*Figure 8.5*). This has enabled calling of minisatellite (O'Dell *et al.*, 2000b) and tetranucleotide and trinucleotide repeat multiallelic microsatellite polymorphisms (Chen *et al.*, 2002). We have been interested in using such sites *within* genes as linkage disequilibrium markers in association studies and wider utility can be anticipated. In our system, microsatellite amplicons are electrophoresed double stranded and there is the possibility that not only length polymorphic information but also internal sequence variation will be accessible by examining heteroduplex mobility.

8.5 Temporal thermal ramp MADGE electrophoresis (Melt-MADGE)

Because MADGE implies a complex spatial configuration of sample wells and tracks, methods involving spatial gradients will be difficult to establish. Our first efforts, however, were focused on the objective of simplifying the use of allele-specific oligonucleotides by determining complete melting profiles rather than single temperature 'snapshots' using real-time thermal ramp electrophoresis (Day *et al.*, 1995b). At this time we had not explicitly recognized that real-time variable-temperature electrophoresis, that is, temporal thermal ramps, would in general be the ideal substitute for spatial chemical gradients for MADGE, for applications involving duplex melting, but this is now clear. This generality has therefore been termed Melt-MADGE, for which we have built prototype apparatus. Two applications have been developed, first, for the determination of complete ASO melting profiles in which the 'freed' rather than the 'bound' ASO is examined (Day *et al.*, 1995b), and second, for *de novo* mutation scanning of complete duplexes, effectively a reconfiguration of DGGE in which the temporal thermal ramp(s) replaces the spatial chemical denaturant gradient as the mechanism of interrogation of full duplex base-pairing. There has recently been a profusion of DGGE variations in the literature, using temperature in space (temperature gradient gel electrophoresis; TGGE) (Henco *et al.*, 1994) and temperature in time (temporal thermal gradient electrophoresis, TTGE) for capillary electrophoresis (Gelfi *et al.*, 1996): these remain inconvenient (generally large apparatus), inherently low throughput (one row of samples across the top of the gel), and expensive (particularly multi-capillary instrumentation). Our system, designated Melt-MADGE, enables analysis of almost 1000 PCR products with 1 h run time in a 2 l tank, using 10 palm-sized MADGE gels. Additional methods under development also utilize spatial gradients in conjunction with MADGE and horizontal gels (Spanakis & Day, unpublished). Nomenclature has become an issue. The term 'gradient' should be reserved for space dependence, 'ramp' for time dependence (LS Lerman, personal communication). We note, furthermore that a 'denaturing gradient' is not necessarily chemical; it can be thermal. Hence TGGE is a variant of DGGE, TTGE is a misnomer. Our gels are horizontal, not vertical. Samples can be loaded either transversely or parallel to the direction of electrophoresis. Mnemonic nomenclature has been adopted in-house, but if systematic nomenclature can be derived, we favour the core term 'melt'.

The apparatus used is purpose-built (Day *et al.*, 1995b). In brief, a $10 \times 10 \times 15$ cm electrophoresis tank contains buffer which is mixed continuously to ensure spatial thermal homogeneity, and the temperature is varied real-time by programmable software controlling heating and cooling systems and receiving feedback from temperature sensors in the tank. Overall, this is little more complex than a PCR machine, although for PCR, low thermal mass for rapid temperature shifts is the objective, whereas for Melt-MADGE, the aim is high thermal mass for stability of small temperature shifts. The initial prototype accommodates 10–12 horizontal MADGE gels spaced in a carrier stack. Each 2 mm thick gel is adherent to and supported by a 2 mm glass plate and is also covered by 2 mm glass, with spacing between each glass/gel/glass sandwich for buffer circulation. Because buffer and hence thermal circulation is vigorous and the glass is both a good thermal conductor and also quite thin, and the thermal ramps are relatively slow and shallow, reasonable spatial thermal homogeneity is achieved, at least to within 0.1°C, as determined by a reference platinum resistivity thermometer.

De novo mutation detection is also an important component of research in genetic disease. For mutations not known in advance, the choice at present is among direct sequencing, *de novo* scanning techniques (Cotton, 1997) to target sequencing more efficiently to appropriate regions, and 'chip' technologies, equivalent with direct resequencing by multiple parallel oligonucleotide hybridization (Chee *et al.*, 1996). For the identification of rare variants, important criteria are the degree of parallelism of samples possible, the number of bases analysed per 'run', and the total number of bases an individual can scan per day. For heterozygote sequencing in humans, the difficulties of accurate base calling (essential for every base to avoid excessive numbers of false positive calls) substantially restrict sequencing as a first line approach (Parker *et al.*, 1996). Chip technologies (Goffeau, 1997) although promising, have low capability for sample parallelism; a chip is expensive ($5000–10 000) for any selected sequence and also requires an optimization process; and the approach will demand high capital expenditure on chip reading hardware.

Scanning technologies are much used, but intense activity to develop improved approaches reflects the importance and shortfalls in this field (Cotton, 1997). Such technologies generally rely on electrophoresis and give some information about the mutation from the resultant band pattern. SSCP appears to be favoured by research laboratories; it is easy to implement as evidenced by 2000 or so Medline references, although sensitivity to all base changes may range from only 30% up to 90% (Sheffield *et al.*, 1993). However, the average sensitivity of SSCP to base changes remains uncertain and the approach in the cited analysis has itself been questioned (Hayashi & Yandell, 1993). DGGE is favoured by diagnostics laboratories: it is laborious to set up but generally more sensitive and robust (Moyret *et al.*, 1994). An objection to DGGE raised by some investigators is the additional cost of oligo clamp synthesis and the occasional adverse influence of the clamp on the PCR reaction itself. The additional cost element corresponds to one or two additional oligos, or to a fluorescence end label which is not necessary in MADGE applications. It is small compared with costs of sequencing many templates, and seems a reasonable compromise to achieve a 'PCR and run' technique evading other intermediate steps and costs. It has not been our experience that GC clamps compromise many PCRs. We have previously reported modifications of SSCP to achieve scanning rates of ≈20 000 bases per day per person (Whittall *et al.*, 1995; Day *et al.*, 1997). However, SSCP has limitations related to the unpredictable effect (if any) of a base change, with resulting inability to predict, control and maximize band resolution, as well as the relative inconvenience of detection systems suitable for single stranded DNA. The folding and tertiary structure and hence electrophoretic mobility of a single strand depends on chance internal base pairing and other influences of bases on overall tertiary structure. The effectiveness is therefore a matter of chance with which a particular single-strand sequence will interrogate all its own internal bases. By contrast, in DGGE the base sequence of one strand is being interrogated in a predictable fashion by the complementary strand.

The principle of DGGE (Fischer & Lerman, 1983) is that strands with a sequence difference or heteroduplex molecules melt and alter their electrophoretic mobility under slightly different conditions of denaturation encountered at some point in space. This is achieved using a gel with either a chemical or thermal gradient from the row of wells of origin down the length of the tracks. However, the use of a spatial gradient has several disadvantages. First, preparing gradient gels is cumbersome and inflexible and sophisticated (non-linear) gradients are practically impossible to prepare. Second, the spatial gradient introduces the constraint of a single, rather than multiple rows or arrays of wells, which severely limits throughput. Finally, the set up does not readily allow the use of very short track lengths, which further reduces the possibility of utilizing high-density arrays of tracks. Melt-MADGE uses the same principle as DGGE or TGGE but exchanges the dimension of the gradient, so that it is a ramp in time rather than a gradient in space. Thus, the PCR products for analysis are loaded on a homogeneous gel and the temperature of the entire gel is raised during the course of the electrophoresis. DNA fragments which form a melted domain at a lower temperature (e.g. heteroduplexes) will display reduced mobility at an earlier stage of the run, and hence will not migrate as far as their more thermostable cognate homoduplexes. A Melt-MADGE proof-of-principle experiment is shown in *Figure 8.6*. The reconfiguration of the denaturing gradient to have temperature as the dependent variable and time as the independent (controlled) variable opens a range of important advantages:

i. It creates the option for high-density arraying of wells and tracks (e.g. MADGE arrays), hence enabling high throughput of analyses.
ii. Programmability of the gradient, which can be absolutely arbitrary and uses temperature rather than chemicals, is achievable easily in time. Arbitrary gradients (rather than simple linear gradients) would be very difficult in space. Gradients as an integral feature of the gel demand special gels. Programmability gives flexibility and convenience.
iii. Potential for arbitrary gradients enables very short track lengths, and short run time (e.g. 1 h) further enables high track density and hence throughput.
iv. The use of ordered arrays (e.g. microplate format of MADGE) much simplifies data handling.
v. Arbitrary programming enables arbitrary resolution, thus reducing the analysis to simple pattern recognition rather than detailed relative mobility measurements. Pattern recognition evades the need for closely juxtaposed tracks or for internal markers. By contrast, the resolution of SSCP is often small, cannot be programmed and hence cannot be arbitrarily great. Thus advantages i–v cannot be realized with SSCP.
vi. For PCR product analysis for sequence variations (known or *de novo*), the system will usually enable the user to 'PCR and run'. This concept represents an operational standard which can be surpassed only by evading electrophoresis altogether, and then only if the liquid phase procedures are simpler, more economical or more sensitive in some way.
vii. Whereas DGGE causes amplicons of similar %(G + C) to migrate to similar positions in the track irrespective of amplicon size, this will not

(a)

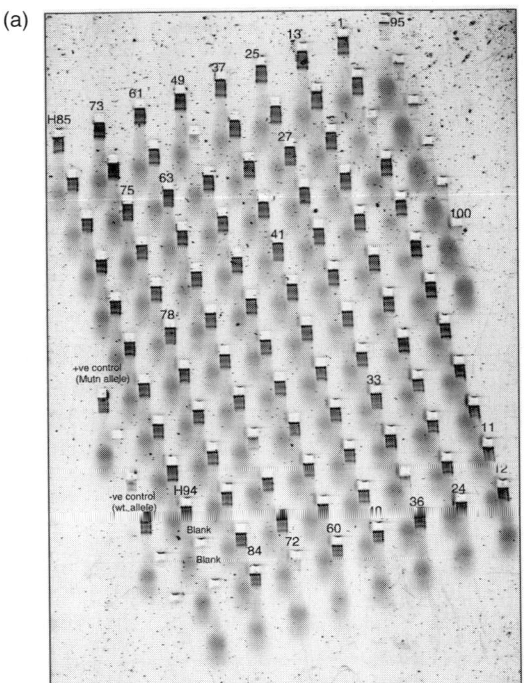

(b)

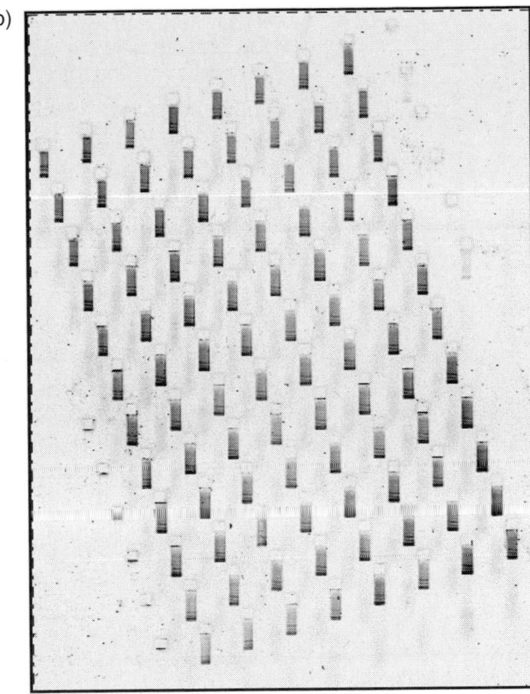

Figure 8.6

Image of a Melt-MADGE analysis. Using a purpose-built temporal thermal ramp electrophoresis apparatus for 'Melt-MADGE,' a DGGE-like analysis of GC-clamped PCR products was undertaken. (a) Scanning of a region of exon 11 (amplicon of GenBank L78833 nucleotides 34042–34340) of the *BRCA1* gene in apparent familial high-risk breast cancer subjects reveals band splitting in some tracks. These subjects have potentially pathogenetic sequence changes but there is also an infrequent single base polymorphism (Q356R) in this amplicon, which probably represents most of the cases (11, 27, 33, 41, 6.3, 75, 78) identified for direct sequence analysis. The scanning technique will identify most if not all single base and small insertion–deletion changes of the amplicon as band splitting, potentially up to four bands in a heterozygote for a mutation, representing the two heteroduplexes and the two homoduplexes. Positive and negative controls are included in 'control' tracks outside the 96. (b) When undertaking population scanning for identification of new mutations, the absence of a natural positive control is a concern. Synthesis of amplicon using a primer with a one base chemical mutation (e.g. at position −4 to −8 relative to the 3′-end) and coannealing with a similar quantity of 'wild type' amplicon (synthesized using primers perfectly matched to the genomic template), gives a suitable artificial positive control. The validation of such a control at every position in the array is shown. Under the temperature conditions used, two resolved heteroduplexes are observed, plus a homoduplex band of twice the intensity (representing co-running wild type and mutant homoduplexes).

occur in the real-time approach, in which migration distance depends heavily on both size and base content. This may facilitate 1D analysis of some amplicon multiplexes more readily.

Melt-MADGE has a resolution theoretically greater than DGGE, as gradients of any complexity can be planned, including, for example, steep initial gradients to resolve heteroduplexes, followed by shallow subsequent gradients to enhance resolution of homoduplexes. Discontinuous gradients to combine analysis of more than one PCR product per run, where their T_m are quite different, will also be feasible. Sequencing can be used as first line analysis. However, parallel sequencing of thousands of samples simultaneously would still be a very substantial and expensive task, and to be successful, no base call in any sequencing reaction should leave any (false) possibility that there could be heterozygosity at that position. In practice, the latter is difficult to achieve (Parker *et al.*, 1996). By contrast, using one strand to interrogate another in a melting assay is an accurate, robust and simple test of bases for perfection (Guldberg & Guttler, 1993). The feasible track usage rate using the first Melt-MADGE prototype is 1000 per 2 h, which would correspond with 20–80 manual or automated instrumentation sequencing gels depending on whether a one track/four label or one track/one label sequencing method were employed. Sample parallelism is also 100–1000 fold greater than with a chip, in which one PCR product can be analysed every 15 min. Additionally, chip techniques require post-PCR procedures such as *in vitro* transcription and fragmentation prior to chip hybridization (Hacia *et al.*, 1996), rather than being PCR-and-run. Sample parallelism is also problematic with capillary electrophoresis instrumentation, although possible at high equipment cost (Clark & Mathies, 1993). The design of a 'Melt-MADGE machine' also means that its final cost is likely to be much less than a chip and 'chip reader': the former requires electrophoresis in equipment otherwise akin to a PCR thermal cycler, the latter requires photolithography, oligonucleotide synthesizer and purpose-built confocal microscope.

These considerations suggest to us that for molecular genetic epidemiological research, MADGE and Melt-MADGE may better meet the criteria of economy, sample parallelism, convenience of set up and accessibility to large and small laboratories alike, and will thus find an important role in future genetic research in complex diseases.

8.6 Software applications (applies to all MADGE approaches)

MADGE image analysis software is often advantageous, as is true for electrophoresis in general. Mobility measurements, band intensities/peak heights, pattern matching, image adjustments, etc. are all feasible and can be systematized to a state of semi-automation. Two features of MADGE to be noted are:

1. Once the grid of lane origins and lane locations are identified by the software (e.g. *Figures 8.3* and *8.5*), other analytical procedures on the lanes are the same as for any other slab gel.
2. There are very many lanes, often with a small number of families of patterns in the lanes. 'One button click' user identification/validation/review of the 'call' for each lane is important for genotyping, as is 'templating' the images such that the act of lane finding by the software is essentially automatic for each new image. These requirements may be unique to MADGE and its applications.

We outline the general principles here, rather than as a specific protocol, in the way that they have usefully applied to different types of MADGE gel image.

8.6.1 Analysing MADGE gels with Phoretix (http://www.phoretix.com) 1D MADGE

8.6.1.1 Mode

Phoretix 1D software has three main modes of analysis, one of which is for MADGE gels. Selecting MADGE mode changes the appropriate dialogues and displays such that the specific requirements for analysing MADGE gels are incorporated or replace the features and facilities for standard gel analysis. This means that all the densitometry, molecular sizing, Rf adjustments and band pattern patching facilities are available.

8.6.1.2 Identifying lanes

Any matrix of x rows and y columns can be identified along with desired lane width. A simple click and drag procedure then fits this matrix to align with the wells. Adjusting the corners of the matrix overcomes any problems associated with gel distortions or poor imaging. A single lane is added anywhere on the gel by click and drag drawing. Once drawn the lane is automatically copied to all other positions using the same angle, length and distance from the well. If necessary, individual lanes may be adjusted for position, angle, width and distortions. All lanes can be identified by number or a letter number combination to keep in line with the original microplate. On good gels, identifying lanes takes no more than a few seconds after the default parameters are set.

8.6.1.3 Background subtraction

Although many MADGE applications do not require precise quantitation, it is advisable to perform background subtraction simply because it can affect band detection as it changes the significance of changes in individual data points. This can be set as an automatic component of analysis in the Analysis Template.

8.6.1.4 Band detection

At this point the gel image changes so that all the lanes are aligned next to each other. Three band detection parameters allow changes to sensitivity, smoothing out noise during band detection and elimination of faint bands. If required, band editing is done with simple mouse clicks on the image or on the data profile graphics.

8.6.1.5 Lane calling

For many MADGE applications, the simplest route to identifying the nature of a lane is to make observational decisions. For this reason the system of lane calling was developed, which allows the use of a computer's numeric keypad to assign names or codes to each lane.

8.6.1.6 Template automation

All the analysis procedures can be saved as an analysis template, which means that dragging a Template onto a gel automates or semi automates future analyses.

8.6.1.7 Flexibility

After band detection all the facilities of Phoretix 1D are available in their standard form including: molecular size determination, Rf adjustments, band pattern matching, relationship dendrograms, wide range of tables and full reporting facilities.

8.7 The future

There is intense activity to systematize biological studies for many organisms, genome-wide clone analysis, genome-wide diversity studies, transcriptome and proteome studies and ensuing functional analyses. In every case, academic as well as commercial, there has been growing emphasis on throughput and complete description, in contrast with conjecture and hypothesis testing. A supplement to *Nature Genetics* 'The Chipping Forecast,' considered this situation in detail (Lander, 1999). There is no doubt that microchip arrays will become major information generators in these programs, using array densities from 10 000 to 10 000 000 grid positions. These systems, however, tend to lack end-user configurability, have restricted range of ready applicability such as nucleic acid hybridizations, require high cost hardware and have required substantial investments from the commercial sector. MADGE, drawing solely on the mature separation technology of electrophoresis which is immediately applicable to many types of analysis, fills a junctional role in medium throughput applications (100–10 000 data points per experiment). There is full end-user configurability and often immediate applicability with minimal development phases and minimal hardware needs. This reflects the origin of MADGE as the electrophoretic equivalent of the industry-standard microplate. We believe that MADGE applications will become very diverse, as have microplate applications. MADGE systems seem likely to occupy a range of niches in biomedical analyses which cannot be fulfilled by 'chips' (assays not feasible or too expensive), where the throughput requirements are moderately high rather than very high, and where basic research requires immediate scale up. In our laboratory the advantages have been mainly for molecular genetic epidemiological studies of human disease but it seems likely that any 96-well (or higher density) PCR block user could benefit from this system. Our research, however, has not explored the worlds of clone, protein, antibody and cell analyses. Application for construction of a high density rat genome map (Watanabe *et al.*, 1999) is a first example in clone analysis. We are aware (personal communications of various researchers) of groups trying to develop MADGE systems for protein, peptide, enzyme and other applications and others developing assay 'services' based around MADGE. Further developments, ranging from basic research to establishment of product and applications to supply the research community, will be essential over the next five years for MADGE to fulfil its role intermediate between 'old-fashioned' slab gels and 'chips' upon which rest many of our future hopes in systematic research in the biological sciences.

Acknowledgements

INMD was a Lister Institute Research Professor during most of these developments. The MRC is thanked for project grants G9605150MB,

G9516890MB and ROPA award. SO'D and BBZ are Wessex Medical Trust Senior Research Fellows. MRJ acknowledges support from the Wellcome Trust. MADGE is subject to patent.

References

Bolla M *et al.* (1995) High-throughput method for determination of apolipoprotein E genotypes with use of restriction digestion analysis by microplate array diagonal gel electrophoresis (MADGE). *Clin Chem* **41** (11): 1599–1604.

Chee MS *et al.* (1996) Accessing genetic information in high-density DNA arrays. *Science* **274**: 610–614.

Chen X *et al.* (2002) Microplate diagonal gel electrophoresis for cohort studies of microsatellite loci. *BioTechniques* in press.

Clark SM, Mathies RA (1993) High-speed parallel separation of DNA restriction fragments using capillary array electrophoresis. *Anal Biochem* **215**: 163–170.

Cotton RGH (1997) *Mutation Detection*. Oxford: Oxford University Press.

Day INM *et al.* (1995a) Dried template DNA, dried PCR oligonucleotides and mailing in 96-well plates: LDL receptor gene mutation screening. *BioTechniques* **18**: 981–984.

Day INM *et al.* (1995b) Electrophoresis for genotyping: temporal thermal gradient gel electrophoresis for profiling of oligonucleotide dissociation. *Nucleic Acids Res* **23**: 2404–2412.

Day INM *et al.* (1997) Spectrum of LDL receptor gene mutations in heterozygous familial hypercholesterolaemia. *Hum Mutat* **10**: 116–127.

Day INM *et al.* (1998) Microplate array diagonal gel electrophoresis (MADGE) systems for molecular genetic epidemiology. *Trends Biotechnol* **16**: 287–290.

Day INM, Humphries SE (1994a) Electrophoresis for genotyping: microtitre array diagonal gel electrophoresis (MADGE) on horizontal polyacrylamide (H-PAGE) gels, Hydrolink or agarose. *Anal Biochem* **222**: 389–395.

Day INM, Humphries SE (1994b) Electrophoresis for genotyping: devices for high throughput using horizontal acrylamide gels (H-PAGE) and microtitre array diagonal gel electrophoresis (MADGE). *Nature* June 36–37 (product review).

Day INM, Wilson DI (2001) Science, medicine, and the future: genetics and cardiovascular risk. *BMJ* **323** (7326): 1409–1412.

Deloukas P *et al.* (1998) A physical map of 30 000 human genes. *Science* **282** (5389): 744–746.

Fischer SG, Lerman LS (1983) DNA fragments differing by single base-pair substitutions are separated in denaturing gradient gels: correspondence with melting theory. *Proc Natl Acad Sci USA* **80**: 1579–1583.

Gelfi C *et al.* (1996) Temperature-programmed capillary electrophoresis for detection of DNA point mutations. *BioTechniques* **21**: 926–928.

Goffeau A (1997) DNA technology: molecular fish on chips. *Nature* **385**: 202–203.

Guldberg P, Guttler F (1993) A simple method for the identification of point mutations using denaturing gradient gel electrophoresis. *Nucleic Acids Res* **21**: 61–62.

Hacia JG *et al.* (1996) Detection of heterozygous mutations in *BRCA1* using high density oligonucleotide arrays and two-colour fluorescence analysis. *Nat Genet* **14**: 441–447.

Hayashi K, Yandell DW (1993) How sensitive is SSCP? *Hum Mutat* **2**: 338–346.

Henco K *et al.* (1994) Temperature gradient gel electrophoresis (TGGE) for the detection of polymorphic DNA and RNA. *Methods Mol Biol* **31**: 211–228.

Kozulic B (1994) Looking at bands from another side. *Anal Biochem* **216**: 253–261.

Lander ES (1999) Array of hope. *Nat Genet* **21** (Suppl. 1): 3–4.

Moyret C *et al.* (1994) Relative efficiency of denaturing gradient gel electrophoresis and single strand conformation polymorphism in the detection of mutations in exons 5 to 8 of the *p53* gene. *Oncogene* **9**: 1739–1743.

Newton CR *et al.* (1989) Analysis of any point mutation in DNA. The amplification refractory mutation system (ARMS). *Nucleic Acids Res* **17**: 2503–2516.

O'Dell S *et al.* (1995) A rapid approach to genotyping of the insertion/deletion polymorphism in intron 16 of the angiotensin converting enzyme gene using simplified DNA preparation and microtitre array diagonal gel electrophoresis. *Br Heart J* **73**: 368–371.

O'Dell SD *et al.* (1996) PCR induction of a *TaqI* restriction site at any CpG dinucleotide using two mismatched primers (CpG-PCR). *Genome Res* **6**: 558–568.

O'Dell SD *et al.* (2000a) SNP genotyping by combination of 192-well MADGE, ARMS and computerized gel image analysis. *BioTechniques* **29**: 500–506.

O'Dell SD *et al.* (2000b). High resolution microplate array diagonal gel electrophoresis: application to a multiallelic minisatellite. *Hum Mutat* **15**: 565–576.

Parker LT *et al.* (1996) AmpliTaq DNA polymerase, FS dye-terminator sequencing: analysis of peak height patterns. *BioTechniques* **21**: 694–699.

Risch N, Merikangas K (1996) The future of genetic studies of complex human diseases. *Science* **273**: 1516–1517.

Schuler GD *et al.* (1996) A gene map of the human genome. *Science* **274** (5287): 540–546.

Sheffield VC *et al.* (1993) The sensitivity of single-strand conformation polymorphism analysis for the detection of single base substitutions. *Genomics* **16**: 325–332.

Studier FW (1973) Analysis of bacteriophage T7 early RNAs and proteins on slab gels. *J Mol Biol* **79**: 237–248.

Sugden B *et al.* (1975) Agarose slab-gel electrophoresis equipment. *Anal Biochem* **68**: 36–46.

Watanabe TK *et al.* (1999) A radiation hybrid map of the rat genome containing 5,255 markers. *Nat Genet* **22**: 27–36.

Whittall R *et al.* (1995) Utilities for high throughput use of the single strand conformational polymorphism method: screening of 791 patients with familial hypercholesterolaemia patients for mutations in exon 3 of the low density lipoprotein receptor gene. *J Med Genet* **32**: 509–515.

Zhang B *et al.* (1999) Functional polymorphism in the regulatory region of gelatinase B gene in relation to severity of coronary atherosclerosis. *Circulation* **99** (14): 1788–1794.

Protocol 8.1: 384-Well MADGE

Products used

PTC-225 DNA Engine Tetrad thermocycler from MJ Research Inc. MADGE equipment now available from Madge-NBS Limited, Huntingdon Science Park, Cambs, UK.

The basic microplate-array diagonal gel electrophoresis (MADGE) system has been adapted to analyse higher density arrays, such as PCR reactions set up in 384-well plates. Here, we present an integrated set of changes to both PCR and MADGE so that 384-well MADGE (MADGE-384) is feasible without robotics and is accessible to any laboratory. MADGE-384 offers a fourfold increase in throughput over conventional gel analysis systems so that now, under almost any conditions (manual, robotic etc.), it is PCR or data calling, not electrophoresis, that limits throughput. As fully automatic software improves, PCR will remain our limiting step.

With 96-well MADGE, the layout of electrophoresed samples matches that in the PCR reaction plate. With MADGE-384, however, PCR reactions are run in four 96-well arrays and the four arrays are intercalated into a single linear format in the MADGE gel (compare *Figure 8.6* and *Figure 8.7*). To avoid any confusion, it is imperative that a precise convention is used for transferring samples and this can be done by clearly defining well positions so that they reflect higher orders of assembly of 96-well arrays. Colour coding is a particularly successful visual aid and *Figure 8.6* shows the well-position nomenclature and colour coding used in our laboratory.

Standardization of PCR reactions for subsequent MADGE analysis

We recommend standardizing assays to the following format.

1. Add DNA solution to wells then dry down by incubating at 80°C for 30 min before adding a PCR 'master mix'.
2. Set PCR reactions up in a final volume of 10 μl.
3. Set assays up in 384-well polypropylene plates (e.g. GRI, 1047-00-0) sealing with adhesive film (Microseal A film, GRI, MSA 5001).
4. Run reactions without oil with a cycling programme that includes a heated lid setting.
5. Use high-performance Peltier thermal cyclers (e.g. PTC-225 DNA Engine TETRAD, MJ Research).

(a)

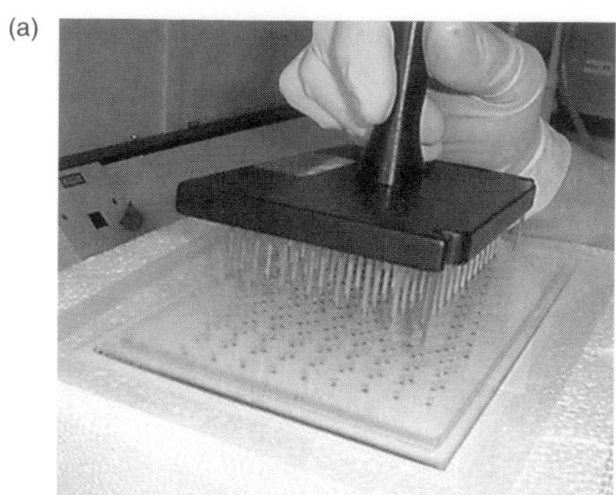

(c)

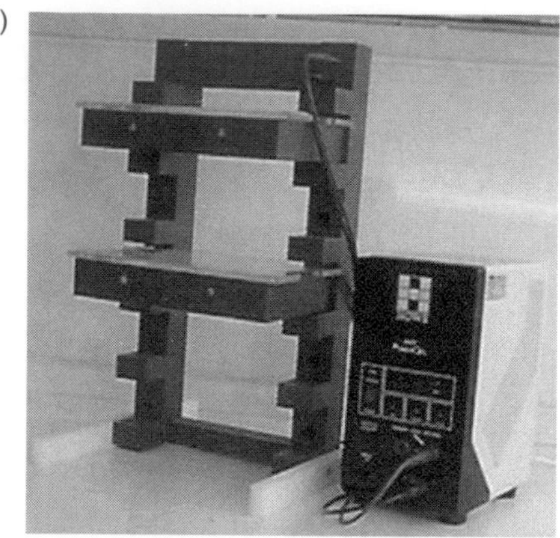

(b)

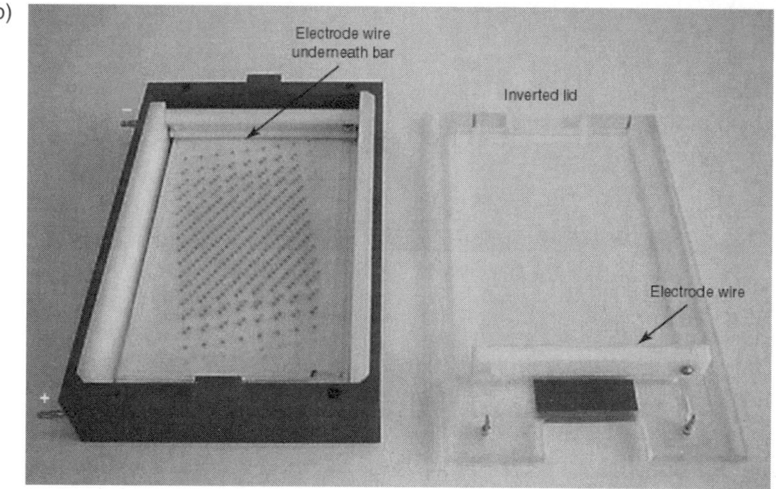

Figure 8.7

Important components of a system for handling 384-well MADGE gels (a) 96-Pin passive replicator loading of 384-well MADGE gels; (b) dry gel box electrophoresis of 384-well MADGE gel; and (c) dry box gel stacking rack for multiple gel electrophoresis.

PCR manual set-up protocol for 384-well plates

Scale up from 96- to 384-well plates requires the adoption of a standardized set-up protocol, to eliminate any chance of user error. To this end, we have devised a simple system to facilitate easy, systematic loading of the 384-well plates. The focus of this protocol is on the MADGE system so there is not the space for these details, but these will be made available on request. Liquid handling robotics for a 384-well PCR set up is the alternative.

Preparing and loading MADGE-384 gels

384 sample PCR reactions can now be analysed directly by loading the samples on a MADGE-384 gel. This system is, in effect, composed of four superimposed 96-well MADGE arrays (*Figure 8.7*).

Table 8.1 Solutions used

Solution	Amount
'Sticky silane'	
Ethanol	495 ml
Glacial acetic acid	2.5 ml
0.5% γ-methacryloxypropyltrimethoxysilane	2.5 ml
10 × TBE	
Tris base	108 g
Boric acid	55 g
0.5 M EDTA (pH 8.0)	40 ml
Deionized water	To a volume of 1 l
MADGE loading dye	
Deionized formamide	980 l
0.5 M EDTA (pH 8.0)	200 l
Xylene cyanol	2.5 mg
Bromophenol blue	2.5 mg
Sterile H_2O	To a volume of 10 ml
25% ammonium persulphate	
Ammonium persulphate	0.25 g
Deionized water	1 ml
Make up every few days and store in the fridge	

A reasonable well size has been maintained (1.5 mm^3) and a track length of 11 mm allows acceptable pattern resolution. The electrophoresis has to be carried out 'dry', to enable 'passive' 96-pin sample transfer. Passive replicators have been chosen to transfer samples in preference to barrel, moving pin or air-displacement systems because they are lighter and more convenient and economical.

1. MADGE-384 gels are poured just as for the 96-well MADGE gels, using the appropriate MADGE-384 former (MadgeBio, M0151), as follows. First, coat one surface of the plain glass plate with 'sticky silane' (99% ethanol, 0.5% glacial acetic acid, 0.5% v/v γ-methacryloxypropyltrimethoxysilane; Sigma, M6514, *Table 8.1*). Use the edge of another glass plate to spread 1 ml sticky silane as evenly as possible. When dry, wipe the surface with a damp tissue.

   MADGE gels can be poured in two different ways. The first is to lay the MADGE gel former horizontal, with the teeth upward, on paper towels or in a flat tray. Ensure the broad border is nearest to you, with the array closest to this short edge. Then mix 35 ml of the required gel mix (*Table 8.2*) and pour this mix into the nearest end of the MADGE gel former (*Figure 8.1b*). Carefully place the glass plate (sticky-silane side down) over the former. It is easiest first to place one end of the plate over the broad edge nearest to you, and then to use one smooth, continuous action to lower the other end down (*Figure 8.1b*).

   The second is to place the glass plate over the former, leaving a small gap next to one of the broad borders. Hold the

Table 8.2 MADGE polyacrylamide gel composition

	Acrylamide content		
	5%	7.5%	10%
Effective separation range (bp)	80–500	70–400	50–300
For 50 ml gel mix:			
10 × TBE	5 ml	5 ml	5 ml
30% acrylamide–bisacrylamide (19:1)	8.3 ml	12.5 ml	16.7 ml
Distilled water	36.7 ml	32.5 ml	28.3 ml
25% ammonium persulphate	150 μl	150 μl	150 μl
TEMED	150 μl	150 μl	150 μl

plates together with one hand and tip the plates to a slight incline so that the gap is at the top end. Carefully pour the gel mix into the gap so that it runs down in between the two plates and fills the mould up from the bottom. This method allows gels to be poured without any acrylamide overspill.

2. The gels are prestained by soaking for 15–20 min in 1 × TBE with 1 μg/ml ethidium bromide.
3. The gel is wiped with a glass rod so that a small residual volume of buffer remains in the well.
4. The PCR plate is spun briefly before removing the sealing film.
5. 5 μl of sample buffer is added to each well using a repeating pipette.
6. The gel is placed on a coloured grid (available at http://www.sgel.humgen.soton.ac.uk) to aid sample loading. The gel is lined up so that the top line of wells lies over the red squares.
7. Samples are transferred from the PCR plate to the MADGE gel using a 96-slot pin replicator (MULTI-BLOT replicator, V&P Scientific, VP408S2) (*Figure 8.7a*). The slot pin replicator is placed over PCR-plate wells in the field corresponding to well 1A1 and 2 μl samples are picked up. The replicator enables the ready location and positioning of the samples but it is important to use no-oil PCR reactions to enable direct pick up of sample.
8. The replicator is then lined up with wells over red squares on the template. The samples are discharged in to the wells by passive diffusion into the residual buffer (*Figure 8.7a*). If the PCR product yield is poor, it is possible to transfer a second, replicate 2 μl sample from PCR plate to MADGE gel.
9. The replicator is blotted clean, rinsed in deionized water and blotted dry. Samples corresponding to well 2A1 on the PCR plate are then removed. The replicator is positioned over the blue squares and the samples discharged.
10. This is repeated with 3A1 field samples going to green squares and 4A1 field samples going to grey squares.

Electrophoresis of MADGE-384 gels

MADGE-384 gels are loaded dry and voltage applied directly across the gel. We have designed a simple gel box to allow safe 'dry' electrophoresis. The MADGE gel is placed at the bottom of the tank and a bar, under which is the cathode electrode wire, is lowered so that it touches the gel (*Figure 8.7b*). Placing the lid onto the tank pushes a bar, under which is the anode electrode, onto the gel, allowing electrophoresis. Running the gel at 150 V for 10 min allows the resolution of most bands in PCR genotyping reactions. We have also designed an electrophoresis stack to allow the simultaneous electrophoresis of up to five gel boxes in parallel (*Figure 8.7c*).

Visualization and analysis of MADGE gels

Given the complexity of the image, MADGE-384 gels have to be analysed using digital imaging systems. The MADGE-384 image can be analysed in 'MADGE mode' of the Phoretix 1D Advanced image analysis software. The software is designed to handle standard 96-well MADGE gels but the 384-well gels can be handled by identifying and analysing each of the four superimposed 96-well arrays separately.

Appendix: Selected Web Resources

http://www.gene-chips.com/
Dr Leming Shi's website provides background and extensive links to review articles, academic laboratories and companies.

http://www.lab-on-a-chip.com/home/index.html
Links to suppliers of microarray equipment and reagents.

http://genome.uc.edu/genome/Web_Resources/arrays.html
Microarray protocol websites.

http://arrayit.com/Products/
Provides protocols for microarray experiments.

http://cmgm.stanford.edu/pbrown/
Dr Patrick Brown's laboratory homepage.

http://genome-www.stanford.edu/microarray
The Stanford Microarray Database (SMD) stores raw and normalized data from microarray experiments, and provides web interfaces for researchers to retrieve, analyse and visualize their data.

http://cmgm.stanford.edu/~schena/index.html
Dr Mark Schena's homepage.

http://www.nhgri.nih.gov/DIR/Microarray/index.html
Microarray Project at the National Human Genome Research Institute.

http://www.ncbi.nlm.nih.gov/ncicgap/
Cancer Genome Anatomy Project at the National Cancer Institute.

http://www.mpiz-koeln.mpg.de/~weisshaa/Adis/DNA-array-links.html
Microarray links on Dr Bernd Weisshaar's homepage.

http://www.mcb.arizona.edu/wardlab/microarray.html
This site has made available results from microarray experiments studying genes that are specifically involved in making worm sperm.

http://www.hgmp.mrc.ac.uk/Research/Microarray/microarrays.jsp
The Microarray Programme of the UK Human Genome Mapping Project Resource Centre.

http://www.hgmp.mrc.ac.uk/GenomeWeb/nuc-genexp.html
A collection of gene expression and microarrays links from UK Human Genome Mapping Project Resource Centre.

http://www.ebi.ac.uk/microarray/index.html
Homepage of the microarray informatics group at the European Bioinformatics Institute.

http://www.mged.org/Workgroups/MIAME/miame.html
Minimum information about a microarray experiment (MIAME) – toward standards for microarray.

http://industry.ebi.ac.uk/~alan/
Dr Alan Robinson's homepage.

http://www.ogt.co.uk/
Oxford Gene Technology Limited founded by Professor Ed Southern.

http://www.affymetrix.com/

http://www.clontech.com/atlas/index.shtml

http://www.resgen.com/intro/microarrays.php3

http://www.incyte.com/index.shtml

http://www.genometrix.com

http://www.genetix.co.uk/

http://www.biorobotics.com/

http://www.cartesiantech.com

http://www.gsilumonics.com

http://www.silicongenetics.com/cgi/SiG.cgi/index.smf
GeneSpring software.

http://www.nhgri.nih.gov/HGP/
Information on the Human Genome Project.

http://www.ensembl.org/

Index